Romana Maria Maier

Antioxidants in seasoning herbs during drought, storage and processing

Romana Maria Maier

Antioxidants in seasoning herbs during drought, storage and processing

Südwestdeutscher Verlag für Hochschulschriften

Imprint

Any brand names and product names mentioned in this book are subject to trademark, brand or patent protection and are trademarks or registered trademarks of their respective holders. The use of brand names, product names, common names, trade names, product descriptions etc. even without a particular marking in this work is in no way to be construed to mean that such names may be regarded as unrestricted in respect of trademark and brand protection legislation and could thus be used by anyone.

Cover image: www.ingimage.com

Publisher:
Südwestdeutscher Verlag für Hochschulschriften
is a trademark of
Dodo Books Indian Ocean Ltd., member of the OmniScriptum S.R.L Publishing group
str. A.Russo 15, of. 61, Chisinau-2068, Republic of Moldova Europe
Printed at: see last page
ISBN: 978-3-8381-2703-3

Zugl. / Approved by: Graz, Karl-Franzens-Universität, Dissertation, 2010

Copyright © Romana Maria Maier
Copyright © 2011 Dodo Books Indian Ocean Ltd., member of the OmniScriptum S.R.L Publishing group

Danksagung

Ich möchte mich sehr herzlich bei Maria Müller für die hervorragende Betreuung und vor allem für die Unterstützung und Motivation zu jeder Zeit und in jeder Lebenslage bedanken.

Weiters möchte ich mich besonders bei meiner Kollegin Martina bedanken, die mich in jeder Phase des Doktoratsstudiums begleitet und unterstützt hat.

Mein Dank gebührt auch Helga Hammer für die Anzucht meiner Versuchspflanzen, Klaus Remele und Volker Wolkinger für die Hilfe bei den HPLCs, meinem Mentor Hartwig Pfeifhofer, Helmut Guttenberger für die Bereitstellung des Arbeitsplatzes und allen Kollegen und Kolleginnen des Institutes für Pflanzenwissenschaften.

Ein herzliches Dankeschön an Bernd Zechmann für die bereichernden Gespräche und an Silvia Haneklaus für die Begutachtung dieser Dissertation.

Ganz besonders möchte ich mich bei meiner Familie und meinem Verlobten Holger bedanken, die mir das Studium ermöglicht und mich zu jeder Zeit unterstützt haben.

Die vorliegende Dissertation wurde unterstützt durch die Bereitstellung des Hordentrockners von Dr. Hans BERGHOLD (Institut für Nachhaltige Techniken und Systeme - chemisch-technische Pflanzennutzung, Joanneum Research Forschungsgesellschaft mbH), von Wetterstationsdaten durch Mag. Dr. Erich PUTZ, Mag. Ing. Helga PIETSCH und Univ.-Prof. Dr. Arnold HANSELMEIER (Geophysik, Astrophysik und Meteorologie (IGAM), Institut für Physik, Karl-Franzens-Universität Graz) und von licht- und elektronenmikroskopischen Bildern durch Johanna SCHÜLLER (Institut für Pflanzenwissenschaften, Karl-Franzens-Universität Graz).

Table of contents

1. Introduction 3
   1.1 The antioxidative defense system in plants (during drought) 3
   1.2 Applied aspects highlighting the use of herbs in food industry 10
2. Aims of these studies 12
3. Material and methods 16
   3.1 Plant descriptions 16
      3.1.1 Nasturtium 16
      3.1.2 Borage 18
      3.1.3 Summer savory 20
   3.2 Plant material 23
   3.3 Meteorological data 23
   3.4 Plant cultivation for experiments of drought stress 24
   3.5 Plant cultivation for field experiments 24
   3.6 Drying and storage experiments 25
   3.7 Processing experiments 26
   3.8 Biochemical analyses 27
      3.8.1 Determination of plastid pigments 27
      3.8.2 Determination of tocopherol (vitamin E) 27
      3.8.3 Determination of ascorbic acid 27
      3.8.4 Determination of glutathione and cysteine 28
   3.9 Determination of photosynthetic parameters 29
4. Drought stress experiments 30
   4.1 Results of nasturtium 30
      4.1.1 Pigments 30
      4.1.2 Tocopherol 33
      4.1.3 Ascorbate 33
      4.1.4 Glutathione and cysteine 34
      4.1.5 Photosynthetic parameters 37
      4.1.6 Summary of the results of nasturtium 41
   4.2 Results of summer savory 42
      4.2.1 Pigments 42
      4.2.2 Tocopherol 45
      4.2.3 Ascorbate 45
      4.2.4 Glutathione 47
      4.2.5 Summary of the results of summer savory 49
   4.3 Results of borage 50

  4.3.1 Pigments *50*
  4.3.2 Tocopherol *53*
  4.3.3 Ascorbate *53*
  4.3.4 Glutathione *55*
  4.3.5 Summary of the results of borage *57*
 4.4 Discussion of the drought stress experiments 58
5. Drying and storage experiments 63
 5.1 Results of nasturtium 63
  5.1.1 Pigments *63*
  5.1.2 Tocopherol *65*
  5.1.3 Ascorbate *65*
  5.1.4 Glutathione *66*
  5.1.5 Summary of the results of nasturtium *67*
 5.2 Results of summer savory 68
  5.2.1 Pigments *68*
  5.2.2 Tocopherol *70*
  5.2.3 Ascorbate *70*
  5.2.4 Glutathione *71*
  5.2.5 Summary of the results of summer savory *72*
 5.3 Results of borage 73
  5.3.1 Pigments *73*
  5.3.2 Tocopherol *75*
  5.3.3 Ascorbate *75*
  5.3.4 Glutathione *76*
  5.3.5 Summary of the results of borage *77*
 5.4 Discussion of drying and storage experiments 78
6. Processing experiments 84
 6.1 Results of nasturtium 84
  6.1.1 Pigments *84*
  6.1.2 Tocopherol *92*
  6.1.3 Ascorbate *96*
  6.1.4 Glutathione *100*
  6.1.5 Summary of the results of nasturtium *107*
 6.2 Discussion of the processing experiments of nasturtium 108
7 Conclusions 113
8. References 116
9. Summary / Zusammenfassung 134
10. Appendix 136

1. Introduction

1.1 The antioxidative defense system in plants (during drought)

Although O_2 is essential for the development of complex life forms on earth, the existence of an organism under these conditions is associated with the generation of reactive oxygen species (ROS). In plants, ROS are formed during different metabolic pathways such as photosynthesis and respiration (ASADA & TAKAHASHI 1987, BARTH & al. 2004) and have important physiological functions (CUI & al. 2004, WILLCOX & al. 2004). For instance, they are not only involved in signaling processes during biotic and abiotic stresses (DESIKAN & al. 2001, NEILL & al. 2002, PUCKETTE & al. 2007) but are also triggering adaptation responses to environmental changes (e.g. drought) (DAT & al. 2000, CRUZ DE CARVALHO 2008, reviewed in FOYER & NOCTOR 2009). On the other hand, ROS often react radically, are in series responsible for the oxidative damage of biological macromolecules like DNA, carbohydrates, lipids and proteins (reviewed in DIPLOCK & al. 1998; LEE & al. 2005) and are affecting cellular functionality (LIMA & al. 2007). Some of the most relevant ROS are: peroxyl radicals (ROO˙), nitric oxide radical (NO˙), super oxide anion radical ($O_2^{˙-}$), singlet oxygen ($^1O_2$) and hydrogen peroxide (H_2O_2). ROS can be either molecules containing at least one unpaired electron (= radical) or reactive non-radical compounds which are also capable of oxidizing biomolecules and are therefore called pro-oxidants, too (DIPLOCK & al. 1998). Normally, the production of ROS is compensated by an elaborate endogenous antioxidant system (LIMA & al. 2007) consisting of two main mechanisms: antioxidant defense with enzymes [like e.g. superoxide dismutase (SOD) and catalase (CAT)] or with non-enzymatic components (e.g. ascorbic acid, tocopherol, glutathione and carotenoids) (*Fig. 1*) (SHAHIDI & al. 1992; RICE-EVANS & al. 1997, reviewed in FOYER & NOCTOR 2009). Therefore, in healthy cells there is an equilibrium between ROS production and defense reactions (MATA & al. 2007). Under optimal conditions, the production of ROS in cells is low, but can increase dramatically due to e.g. unfavorable environmental conditions (POLLE 2001, MITTLER & al. 2004) like high light intensities, heavy metals, drought or pathogen attacks.

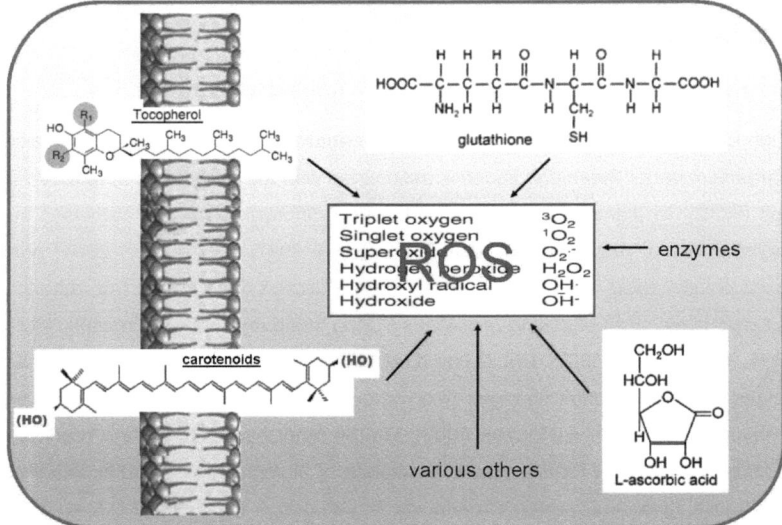

Figure 1. Scheme showing some of the involved components of the plant defense network detoxifying reactive oxygen species (ROS). Lipid-soluble carotenoids and tocopherol are membrane bound (various authors, see above).

However, abiotic stresses trigger a wide variety of plant responses ranging from altered gene expression to changes in cellular metabolism and growth. These responses can be affected by duration and magnitude of the stress, plant genotype and its developmental stage and whether or not the stress factors act alone or in concert (KOZLOWSKI & PALLARDY 2002).

As the availability of water is one of the major factors limiting the plant productivity (NEUMANN 2008), this applied work has focused on stress induced by drought. Most of the damaging effects of **drought** in plants are associated with the photosynthetic process. To avoid drought plants can either minimize water loss (e.g. stomata closure, small leaves, etc.) or maximize water uptake (e.g. tapping ground water by deep roots) (KOZLOWSKI & PALLARDY 2002). Drought resistant plants, for instance, show reduced leaf and stem growth rates (MUNNS & SHARP 1993), maintenance or increase in root extension rates (HOSE & al. 2000, PARENT & al. 2009) or alternative additional drought avoidance/tolerance strategies like e.g. induction of senescence or synthesis of osmotically active solutes (CHAVES & al. 2003). As suggested in many studies, one of the early plant responses to drought may be the increase of abscisic acid (ABA), inducing stomata closure to reduce leaf transpiration (SCHROEDER & al. 2001, KOZLOWSKI & PALLARDY 2002, LUAN 2002, reviewed in FOYER & NOCTOR 2009). On the other hand, stomata closure causes the limitation of CO_2 fixation and in series, an excess of excitation of energy is not dissipated which leads to the production of ROS (ASADA 1999, CHAVES & OLIVIERA 2004, GRASSI & MAGNANI 2005). However, under stress conditions a combination of antioxidants plays an important role to detoxify ROS in order to protect plants against peroxidation, e.g. in maintaining the integrity of photosynthetic membranes (HAVAUX 1998; reviewed in NOCTOR & FOYER 1998 and SMIRNOFF & WHEELER 2000; MUNNÉ-BOSCH & ALEGRE 2002, MUNNÉ-BOSCH 2005, ASADA 2006). According to the literature, the concentrations of antioxidants can show increases, decreases or no effects (as drought stress response) and are depending on plant species as well as the intensity and duration of stress (ZHANG & KIRKHAM 1996, BOO & JUNG 1999, EGERT & TEVINI 2002, JUNG 2004, MUNNÉ-BOSCH & PENUELAS 2004, LEI & al. 2006, MÜLLER & al. 2006, GALMES & al. 2007a and b).

At present, a common theme dealing with stress responses is the **phenomenon of priming**, which shows that previous exposure to stress can make a plant more resistant to future exposure (reviewed in BRUCE & al. 2007). In many crop species, this effect is (long) known as drought acclimation or drought hardening (SELOTE & KHANNA-CHOPRA 2006, SRIVALLI & al. 2003). Pre-stressed plants can activate their plant defence network either faster, and/or stronger. Therefore, there seems to be a mechanism in plants for storing information ("memory") and evidence that plants are adapted to metabolic modifications (BRUCE & al. 2007).

In general (independent of the type of stress), the main non-enzymatic antioxidants in (all) plants are carotenoids, tocopherol, ascorbate and glutathione (NOCTOR & FOYER 1998, HALLIWELL & GUTTERIDGE 1999).

Carotenoids - lipophylic isoprenoid compounds (BOTELLA-PAVÍA & RODRÍGUEZ-CONCEPCIÓN 2006) - are naturally occurring plant pigments of yellow, orange and red spectrum (MAHER 2010) and therefore found in many flowers, fruits and vegetables. There are two classes of carotenoids in nature: carotenes (e.g. α-carotene, β-carotene) and their oxygenated derivatives called xanthophylls (e.g. lutein, violaxanthin, antheraxanthin, neoxanthin and zeaxanthin) (BOTELLA-PAVÍA & RODRÍGUEZ-CONCEPCIÓN 2006). They are mainly synthesized *de novo* in photosynthetic organisms (MADSEN & al. 1998) and are integral components of the photosynthetic apparatus (HUMBECK & al. 1989) embedded in functional pigment-binding protein structures of the thylakoid membranes. Carotenoids are acting as accessory light harvesting pigments and have photo-protective functions (e.g. quenching of chlorophyll triplets and in series preventing generation of singlet oxygen) (BOTELLA-PAVÍA & RODRÍGUEZ-CONCEPCIÓN 2006). They also serve as precursors for plant hormones, abscisic acid and strigolactones (reviewed in CAZZONELLI & POGSON 2010). Especially, the xanthophyll cycle has two important main functions: thermal dissipation and detoxification of ROS. It is a light-dependent cyclic conversion between three xanthophylls (YAMAMOTO 1979, DEMMING-ADAMS & ADAMS 1992). After increase of the photon flux density, the acidification of thylakoid lumen (= low pH) activates violaxanthin de-epoxidase, which converts violaxanthin via antheraxanthin to zeaxanthin and when energy dissipation is no longer required violaxanthin is re-synthesized by zeaxanthin epoxidase (HAGER 1969, HAGER 1975, HAGER & HOLOCHER 1994, MÜLLER-MOULE & al. 2002, GROUNEVA & al. 2006) (*Fig. 2*). Animals, including humans, cannot synthesize carotenoids and therefore rely on the diet as source. Being the precursors of vitamin A, carotenoids provide health benefits based on their antioxidant properties (e.g. scavenging of ROS singlet oxygen and peroxyl radicals) (DIPLOCK & al. 1998, BOTELLA-PAVÍA & RODRÍGUEZ-CONCEPCIÓN 2006). Overlapping functions of antioxidants (allowing one to substitute for another) are suggested not only by the fact that singlet oxygen is scavenge by β-carotene at low and α-tocopherol at high oxygen partial pressure, but also by the accumulation of zeaxanthin when tocopherol is lacking and vice versa (*figure 4*) (BURTON & INGOLD 1984, HAVAUX & al. 2005, FRANKEL 2007).

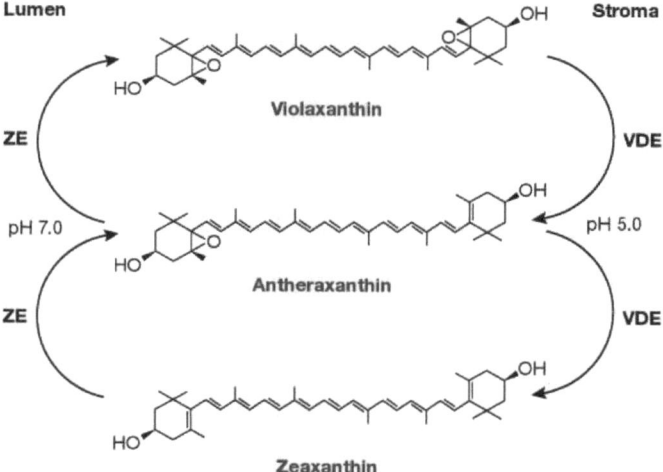

Figure 2. The xanthophyll cycle. Low pH activates violaxanthin de-epoxidase (VDE), converting violaxanthin via antheraxanthin to zeaxanthin. Violaxanthin is re-synthesized by zeaxanthin epoxidase (ZE) (SZABÓ & al. 2005).

Tocopherols are lipid-soluble molecules belonging to the group of vitamin E compounds. They consist of a chromanol ring system and a saturated polyprenyl side chain, can be subdivided in α-, β-, γ- and δ-tocopherol and are found in all photosynthetic organisms (reviewed in FALK & MUNNÉ-BOSCH 2010). Being highly lipophilic, they are operative in membranes (e.g. cell membrane or intra- and extracellulary) (RODRIGO & al. 2007) (*Fig. 1*) and lipoproteins (DIPLOCK & al. 1998). In photosynthetic membranes α-tocopherol reduces ROS levels and in series limits the extent of lipid peroxidation (TREBST & al. 2002, HAVAUX & al. 2005, MAEDA & al. 2005). To prevent or delay oxidative deterioration and increase food quality (COSIO & al. 2006), α-tocopherol is used as a food additive (DIPLOCK 1992, LANGSETH 1995). In comparison with other lipophlic antioxidants, α-tocopherol is probably the most efficient in the lipid phase (DIPLOCK & al. 1998) and meets human vitamin E requirements best (TRABER & ATKINSON 2007) due to its highest bioavailability (RODRIGO & al. 2007) and the highest vitamin E activity (reviewed in DELLAPENNA & LAST 2006). After reacting as antioxidant the tocopheroxyl radical is reduced by glutathione or ascorbate (*Fig. 4*) (DIPLOCK & al. 1998). In single-gene transgenic lines the accumulation of vitamin E resulted in changes of the ascorbate pool (LI & al. 2010) assuming complex connections.

Ascorbate (vitamin C) is well examined because of both, its importance for human health and as one of the major antioxidants in plant species. It is water soluble, quantitatively the predominant antioxidant in plant cells and is found in all subcellular compartments, the cytoplasma as well as the apoplast (see e.g. JIMENEZ & al. 1997 and 1998, VANACKER & al. 1998, SMIRNOFF 2000, KOLLIST & al. 2001, VAN HOVE & al. 2001, NOCTOR & al. 2002, TAKAHAMA 2004). Under optimal physiological conditions ascorbate exists in leaves (which often contain more ascorbate than chlorophyll) mostly in the reduced form (90%) and the ascorbate pool represents 10 % of the soluble carbohydrates (NOCTOR & FOYER 1998, BLOKHINA & al. 2003). Ascorbic acid oxidizes readily in aqueous solutions into monodehydro-ascorbic (MDHA) and dehydro-ascorbic acid (DHA). These forms contribute to the biological activities of vitamin C (ULRICH-MERZENICH & al. 2009), which is considered to be one of the most powerful, least toxic natural antioxidants acting as a scavenger of ROS especially against the superoxide radical anion, H_2O_2, the hydroxyl radical and singlet oxygen. However, ascorbate might (in combination with free transition metal ions like Fe and Cu) act as a pro-oxidant in vivo, too. Besides its function as antioxidant, vitamin C is also active as co-factor for several important enzymes (DIPLOCK & al. 1998) and is involved in redox signaling and modulation of gene expression (NOCTOR & FOYER 1998; NOCTOR 2006, FOYER & NOCTOR 2009). Plants provide the major source of dietary vitamin C for humans (reviewed in SMIRNOFF & al. 2001) and especially fruits and vegetables are good sources. Embedded in the plant metabolism, ascorbate is not only able to regenerate tocopherol (DIPLOCK & al. 1998) and to act as co-factor for violaxanthin de-epoxidase (SMIRNOFF & al. 2001) but is also an important part of the **ascorbate-glutathione-cycle:** Ascorbate is oxidised to monodehydro-ascorbate (MDHA) by ascorbate peroxidase, reducing H_2O_2 to H_2O. When the originated MDHA is reduced back to ascorbate (by monodehydroascorbate reductase (MDHAR) using NAD(P)H) also some dehydroascorbate (DHA) is produced. By action of DHA-reductase DHA is reduced to ascorbate using GSH which is oxidized to GSSG (the latter can be reduced back to GSH by glutathione reductase using NADPH) (NOCTOR & FOYER 1998, FOYER & NOCTOR 2009) (*Fig. 3, 4*).

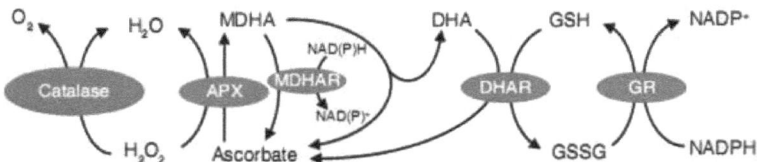

Figure 3. The **ascorbate-glutathione-cycle.** APX = ascorbate peroxidase, DHA(R) = dehydroascorbate(reductase), GR = glutathione reductase, GSH = reduced glutathione, GSSG = oxidized glutathione, MDHA(R) = monodehydroascorbate-(reductase) (FOYER & NOCTOR 2009).

Glutathione is a major water-soluble antioxidant in plant cells reducing directly most ROS. It is found in the vast majority of both, pro- and eukaryotic cells and is an important storage form of reduced sulphur (in many cells more than 90% of the total nonprotein sulphur!). The reduced form of glutathione (GSH) is a tripeptide thiol consisting of glutamate, cysteine and glycine (oxidized form = glutathione disulphide GSSG) (MEISTER 1988, NOCTOR & al. 1998, FOYER & NOCTOR 2009). Besides its function as antioxidant and important part of the ascorbate-glutathione-cycle, it acts as regulator of gene expression, is precursor of phytochelatins, substrate for the GSH S-transferases (catalyzing the conjugation of GSH with potentially dangerous xenobiotics e.g. herbicides) and is involved in stress signalling and redox regulation of the cell cycle (NOCTOR & al. 1998; FOYER & al. 2001, EDWARDS & al. 2005, DERIDDER & GOLDSBROUGH 2006, MAUGHAN & FOYER 2006 and FOYER & NOCTOR 2009; SZALAI & al. 2009). Furthermore it can protect proteins from oxidation by a process called glutathionylation (HURD & al. 2005 a, b) and might even be involved in the activation of cell death (FOYER & al. 2001). However, glutathione is a vital intracellular and extracellular protective antioxidant against oxidative/nitrosative stresses, which play a key role in the control of many human diseases, too (MAUGHAN & FOYER 2006).

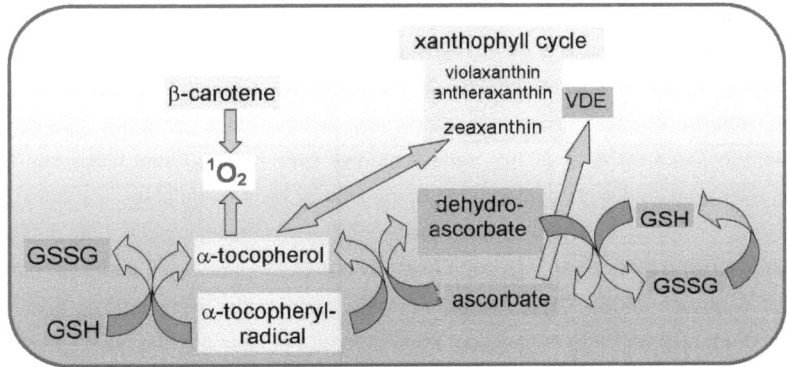

Figure 4. Scheme showing connections between some of the involved substances. $^1O_2$ = singlet oxygen, GSH = reduced glutathione, GSSG = oxidized glutathione, VDE = violaxanthin-de-epoxidase, yellow = lipid-soluble, blue = water-soluble (various authors, see above).

Due to the fact, that all substances mentioned above are important for plants and humans, various papers, books and reviews deal with their biosynthesis, metabolism and biotechnological modifications. Therefore general information about antioxidants can be found in the review of BLOKHINA & al. (2003), the review of DIPLOCK & al. (1998) and the book of FRANKEL (2007) dealing with antioxidants in food and biology as well as the review of ZUSSMANN & al. (2010), dealing with medical aspects of vitamins. To get a good overview on carotenoids the recent reviews of BOTELLA-PAVÍA & RODRÍGUEZ-CONCEPCIÓN (2006), LU & LI (2008) and FARRÉ & al. (2010) are recommended as well as the reviews about tocopherol of DELLAPENNA & LAST (2006), TRABER & ATKINSON (2007) and FALK & MUNNÉ-BOSCH (2010). Excellent literature dealing with ascorbate and/or glutathione are the reviews of NOCTOR & FOYER (1998), NOCTOR & al. (1998), DAVEY & al. (2000), SMIRNOFF & al. (2001), MAUGHAN & FOYER (2006) and FOYER & NOCTOR (2009). Exemplarily, for the interesting field of "stressful memories" the review of BRUCE & al. (2007) is recommended.

1.2 Applied aspects highlighting the use of herbs in food industry

Aromatic herbs and spices have been collected, grown and added to different types of food since ancient times worldwide. This is not only due to their organoleptic properties and to improve the flavor (reviewed in DIPLOCK & al. 1998), but also to avoid deterioration (MATA & al. 2007, TAJKARIMI & al. 2010). At present, there is an increasing interest of both, the (food and pharmaceutical) industry and the scientific research in spices and aromatic herbs because of their strong antioxidant and antimicrobial properties (SUHAJ 2006, reviewed in TAJKARIMI & al. 2010). Various studies even point out that herbs can have a higher antioxidative activity compared to berries, other fruits, vegetables and nuts (ZHENG & WANG 2001, WOJDYLO & al. 2007). Furthermore, the increasing supply of processed food (by food industry) leads to an increased demand and use of seasoning herbs (BOELT 1990, KAEFER & MILNER 2008). These herbs enrich the main dishes with vitamins and mineral salts and are also affecting the sensory traits of food. For example, the color (which is one of the most important sensory traits) is improved by plant pigments (LISIEWSKA & al. 2004; BOTELLA-PAVÍA & RODRÍGUEZ-CONCEPCIÓN 2006) and therefore, the latter have been exploited industrially for a long time. For instance, various carotenoids are employed to provide the typical color of e.g. salmon, shrimp, eggs, poultry or margarine (BOTELLA-PAVÍA & RODRÍGUEZ-CONCEPCIÓN 2006). To prevent (natural) spoilage processes in food (caused by micro-organisms), spices and herbs are added because of their antimicrobial effects on plant and human pathogens (TAJKARIMI & al. 2010). However, another major cause of food quality deterioration is lipid oxidation (LABUZA 1996) which results (beside the growth of undesirable micro-organisms) in the development of spoilage, off-flavor, rancidity, and deterioration, making such products unacceptable for human consumption (BOZIN & al. 2007). Therefore, one of the highest priorities of the food industry has been the control of oxidative processes (DIPLOCK 1998) and a number of food additives and chemical products have been used for food preservation (DZIEZAK 1986). Although some synthetic antioxidants are very efficient e.g. in preventing auto-oxidation (CAILLET & al. 2007), nowadays their use in food is restricted due to safety concerns and reports on their involvement in chronic diseases (CEBALLOS & FERNANDEZ 2000). Some synthetic antioxidants were even found to be toxic and carcinogenic in animal models (ITO & al. 1986, SAFER & AL-NUGHAMISH 1999) and only a few compounds are currently approved for use in the food industry (CAILLET & al. 2007). As a consequence, there is a world-wide trend to use natural additives in food and cosmetics and attention is directed towards natural antioxidants mainly from plant sources (e.g. CEBALLOS & FERNANDEZ 2000, GÜLÇIN

& al. 2002, OKTAY & al. 2003). To search for these natural antioxidants (besides fruits and vegetables), herbs and spices are one of the most important targets (reviewed in YANISHLIEVA & al. 2006). In the emerging nutrition industry herbal medicines have a great potential being both, medicines and foods (HÄNSEL & HAAS 1984) and embraced by the terms "nutraceuticals" or "functional food", they had a big impact on the food industry (reviewed in KAUR & KAPOOR 2001).

However, for industrial use it must be kept in mind that concentrations of plant originated antioxidants can be effected by various factors not only occurring during cultivation and production (KALT & al. 1999) but also during food processing. For instance, thermal treatments are very common in food industry, e.g. blanching is one of the pre-treatments used before freezing (MOUNTNEY & GOULD 1988). To reduce the moisture content of the plant material various drying methods are applied assuring that there is no substantial loss of flavor, taste, color or nutrients (ARSLAN & ÖZCAN 2010). In general, it is well known that e.g. temperature, pH of the media, processing treatment (e.g. heat) and storage can strongly influence the activity and contents of antioxidants (GAZZANI & al. 1998). Especially water-soluble antioxidants like ascorbic acid are sensitive to water loss and in general, various studies (e.g. NEGI & ROY 2004, DAOOD & al. 2006) showed, that the degradation of antioxidants is often following first order kinetics depending on temperature and (storage) duration (DAOOD & al. 2006, LAVELLI & al. 2006, GUTZEIT & al. 2008, HIDALGO & al. 2009, SABLIOV & al. 2009). Furthermore, e.g. pigment stability varies between different food even if the same storage/processing conditions are used and in series, optimum conditions for preparation, storage etc. differ, too (RODRIGUEZ-AMAYA 2009) and must be identified for each herb. The concentrations and synergies of various substances are responsible for the antioxidative properties of herbs/foods, but even if e.g. a product lacks antioxidant properties, it may still possess other important biological properties (CAILLET et al. 2007).

2. Aims of these studies

Drought stress experiments:
It is well known, that stress influences the plant metabolism and concentrations of substances or enzymes activities (belonging to the plant defense network) and e.g. photosynthetic processes are modified, whereas especially the stress intensity is essential. Therefore, in our studies climate chamber grown plants of all three plant species are exposed to a mild drought stress which assures that the plant defense network is activated but there is no damage in the plants. Concentrations and ratios of all substances and photosynthetic parameters (mentioned above) are measured at plants without stress (= control), after mild drought stress and after re-watering. Our studies do not just focus on the metabolism during stress but in particular on the status after and therefore, they allow us to examine changes in the concentrations of antioxidants and the photosynthetic parameters during stress and especially after recovery because it is assumed, that plant metabolism is modified during stress and that this information can be "memorized" by the plant allowing her do react faster and/or stronger to further stress attacks.

The following questions shall be answered:

- *How do stress and recovery influence the metabolism of seasoning herbs?*
- *Are there differences between the plant species?*
- *Are plants "memorizing" the stress by adapting their metabolism?*
- *Can mild drought stress be used to prime the plants to be better prepared for further stress attacks?*
- *Are the examined plant species good antioxidant sources?*
- *Is mild drought stress a possibility to enhance antioxidant contents in plant material for e.g. (food) industry?*

Drying and storage experiments:

Many studies show that antioxidants are sensitive to e.g. water loss, light or heat and can degrade easily which is a major problem for (food) industry. However, in plants water-soluble (e.g. ascorbate and glutathione) and lipid-soluble (e.g. carotenes, tocopherol) antioxidants can be found and according to their chemical properties they are differently sensible to degradation. Also the concentrations and compositions of them in plants are species-dependent. In our studies field grown plant material of all three species is dried in a hurdle ensuring constant temperature of 35 °C, good ventilation and protection from light. After drying the material is stored for 3 respectively 6 months in the dark, with constant room temperature and in two different types of bags (paper bags, and tie bags equipped with foil and used e.g. for herbal teas) which are permeable to air. Thereafter, antioxidant contents are examined after drying and storage and compared with fresh controls to find out if and how massive the concentrations of antioxidants change during these processes.

The following questions shall be answered:

- *Are the antioxidants degrading during drying and storage in plant material and if yes, how massive?*
- *Do differences between antioxidants exist?*
- *Are there even antioxidants left after drying and storage for months?*
- *Is there a difference between storage in paper bags and tie bags?*
- *Is there a difference between the plant species?*
- *Is one of the examined species a good antioxidant source even after drying and storage and if yes, for what antioxidants?*

Processing experiments:

Antioxidants are (due to their chemical properties) not only sensitive to drying and storage but also to different processing methods. However, processing is quite common in industrial food production and sometimes treatments like blanching are even adopted to enhance food quality. Furthermore, herbs are often added to food not only fresh but also dried and in different processing stages to e.g. increase flavor. In our studies we simulate processing and cooking at home. Therefore, dried plant material as well as fresh leaves of field grown nasturtium are used and processed in different ways. Dried material is boiled for 5 and 20 minutes, fresh material is cut and immediately boiled for 5 and 20 minutes or first cut, let rested and then boiled for 5 and 20 minutes. In series, contents of the substances (mention above) are examined and compared with the dried and fresh controls, respectively. As herbs are not only good sources for essential oils (flavoring and preserving food) but also for other antioxidants we examine in which of the examined processing methods antioxidants (mentioned above) are still left and which extent.

The following questions shall be answered:

- *Does the processing of the plant material lead to a loss of antioxidants and if yes, how massive?*
- *Is nasturtium a good antioxidant source even after processing and if yes, for what antioxidants?*
- *Is it advisable to add (fresh or dried) herbs as antioxidant source to food?*
- *How does the processing method or duration influence the degradation of antioxidants?*
- *In which point of time is it advisable to add herbs (as antioxidant sources)?*
- *Can antioxidants even be found in dried and processed plant material?*

Aims

To answer all these questions in three important seasoning herbs (*Tropaeolum majus* **L.**, *Borago officinalis* **L.** and *Satureja hortensis* **L.**) the concentrations of

- chlorophyll a and b
- α- and β-carotene
- lutein plus zeaxanthin
- violaxanthin
- antheraxanthin
- neoxanthin
- α-tocopherol
- ascorbate (reduced/oxidized)
- glutathione (reduced/oxidized)

are examined. Furthermore these photosynthetic parameters are measured:

- net-photosynthesis
- transpiration rate
- stomatal conductance
- intracellular CO_2 concentration
- light response curves
- CO_2 response curves

Experiments are divided into three different main parts:

- Drought stress experiments
- Drying and storage experiments
- Processing experiments

Our results shall give important information for the science community about the plant defense network and metabolism of the examined three seasoning herbs, the changes during mild drought stress and recovery, a possibly "stress memory effect" as well as the degradation of antioxidants during drying, storage and processing in plant material. Especially for food producers our studies could demonstrate not only the possibility of active enhancing antioxidants in seasoning herbs using mild drought stress, but also the influence of drying, storage and processing on the plant material. In series, the implementation of our results by food industry could lead to increased food quality for end consumers.

3. Material and methods

3.1 Plant descriptions

3.1.1 Nasturtium

Nasturtium (also called indian cress - *Tropaeolum majus* L.) belongs to the family of Tropaeolaceae and the order of Brassicales (which includes several food relevant families) (SCHREINER & al. 2009). It is an annual creeping plant flowering from red to yellow (CHEVALLIER 2000) (*Fig. 5 A – C*). The plant is called "nasturtium" as an allusion to its spicy flavour (in Latin: nasum = nose and torquere = to grimace) (SCHULTZ & GMELIN 1954). Nearly all organs (inflorescences, leaves, and unripe green seeds) of nasturtium are edible and consumed as herbal plant products, e.g. inflorescences and leaves are used in salads, vinaigrettes and sauces; green seeds are pickled and used as substitute for capers (NIIZU & RODRIGUEZ-AMAYA 2005).

Furthermore nasturtium is containing high vitamin concentrations (especially ascorbate), glucosinolates (responsible for its peppery flavour), high lutein contents and mustard oils and therefore various therapeutic properties have been reported (LYKKESFELDT & MOLLER 1993, SULZBERGER 2002, MÜLLER 2004, TORRES-JIMENEZ & QUINTANA-CARDENES 2004, NIIZU & RODRIGUEZ-AMAYA 2005). Phytochemical screening showed the presence of fatty acids (erucic acid, oleic acid, linoleic acid), benzyl isothiocyanate, and flavonoids (isoquercitroside, quercetol 3-triglucoside, and kaempferol glucoside) in the seeds and leaves of *Tropaeolum majus* L. (DE MEDEIROS & al. 2000; MIETKIEWSKA & al. 2004, ZANETTI & al. 2004). Nasturtium is characterized by a high concentration of the aromatic glucosinolate glucotropaeolin, which was quantitatively determined in inflorescences, leaves, and unripe green seeds (e.g. BLOEM & al. 2007, SCHREINER & al. 2009). But not only glucotropaeolin, also sinalbin and tetracyclic triterpene cucurbitacins have been isolated from the leaves of this plant (LYKKESFELDT & MOLLER 1993, GRIFFITHS & al. 2001). Recently, aromatic isothiocyanates derived from gluconasturtiin and glucotropaeolin are intensively discussed to be strongly anticarcinogenic (e.g. STICHA & al. 2002, MIYOSHI & al. 2004) generating considerable nutritional and pharmacological interests (TALALAY & FAHEY 2001, HOLST & WILLIAMSON 2004). Additionally, nasturtium has relative high concentrations of phenolic compounds (SANTO & al. 2007). As described in NIIZU & RODRIGUEZ-AMAYA (2005) especially flowers and leaves of nasturtium are an excellent food source of lutein.

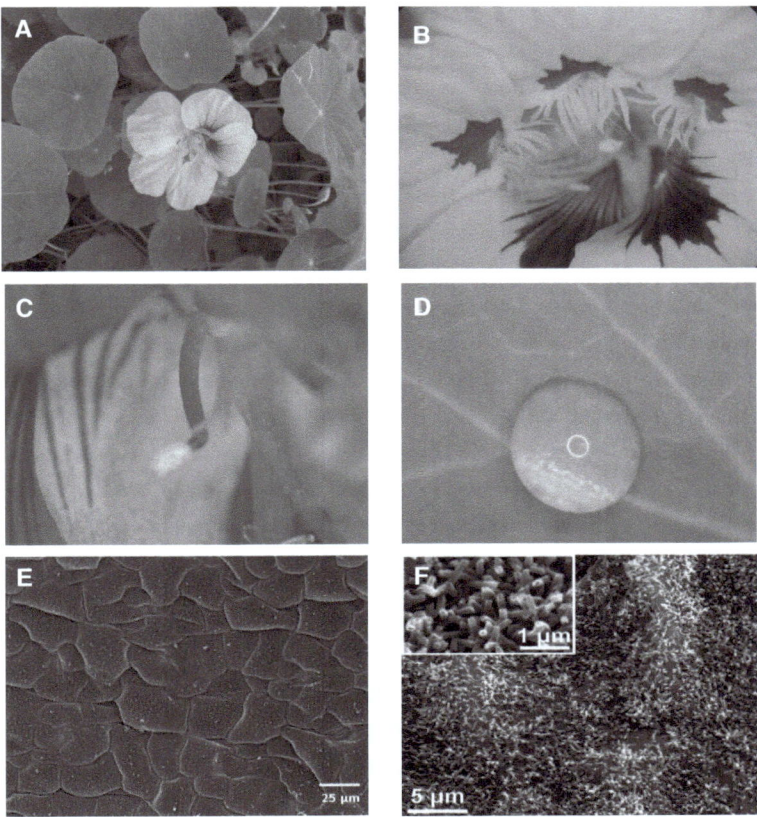

Figure 5. *Tropaeolum majus* L. (nasturtium) **A)** Plant field-grown and flowering. Macroscopic image of **B)** a yellow flower **C)** an orange flower. **D)** Water droplet on the leaf surface "Lotus–effect". **E)** SEM images of the leaf surface (SCHÜLLER 2010) **F)** SEM image of the wax tubules on the upper leaf surface, inset shows detail (KOCH & al. 2006) (for magnitudes see scale bars).

Content and composition of anthocyanins were unknown for long time and described tentatively for nasturtium first in HARBORNE (1963), who showed, that the main pigment present in the orange petals is pelargonidin-3-sophoroside. Analyses of nasturtium have also led to the identification of two new families of odorant compounds, thiocarbamates and thiocarbonates (BREME & al. 2007 and 2009). As there is the trend of replacing synthetic colorants in cosmetics, food and pharmaceuticals, nowadays, nasturtium plants have become a potential natural source. The data of HARBORNE (1963) were confirmed by GARZÓN & WROLSTAD (2009) who also showed high antioxidative activities in nasturtium with values even higher than in blue-berries or *Rubus* species and close to that presented for cranberries and strawberries (KATSUBE & al. 2003). *Tropaeolum majus* L. is also an important medicinal plant widely distributed in the world and especially in Brazilian folk medicine, the leaves are often used in the treatment of several disease including cardiovascular disorders, urinary tract infections, asthma, and constipation (CORRÊA 1978, FERREIRA & al. 2004, FERRO 2006). Therefore, various pharmacological experimental studies have been carried out with *Tropaeolum majus* L. (e.g. BINET 1964, PICCIARELLI & al. 1984, PICCIARELLI & ALPI 1987, PINTÃO & al. 1995, DE MEDEIROS & al. 2000, GOOS & al. 2006, GASPAROTTO & al. 2009).

Nasturtium leaves are covered with tubular wax chrystals containing the secondary alcohol nonacosan-10-ol and nonacosanediols as main compounds and various other components (various authors - published in KOCH & al. 2006). The wax tubules are three-dimensional and provide hydrophobicity and surface roughness, responsible for the superhydrophobicity of the leaves. Therefore, the "lotus effect" can be observed. The epicuticular waxes are responsible for several other properties, such as visible light reflection and UV-absorption, reduction of insect, particle and pathogen adhesion (*Fig. 5 D – F*) (various authors – published in NIEMITZ & al. 2009).

3.1.2 Borage

Borage (*Borago officinalis* L.), a Boraginaceae, is an annual herbaceous plant, flowering in various shades of blue (CHEVALLIER 2000) (F*ig. 6 A – C*). It originated in the Mediterranean region but is nowadays cultivated worldwide (KAPOOR & NAIR 2005). Therefore it is native to Europe, North Africa, and Asia Minor (BEAUBAIRE & SIMON 1987) and grows wild in Central and Eastern Europe (KATZER 1999). It needs soils which are rich in humus (DACHLER & PELZMANN 1999) and is an important vegetable crop cultivated in some countries (KAPOOR & NAIR 2005) including Iran.

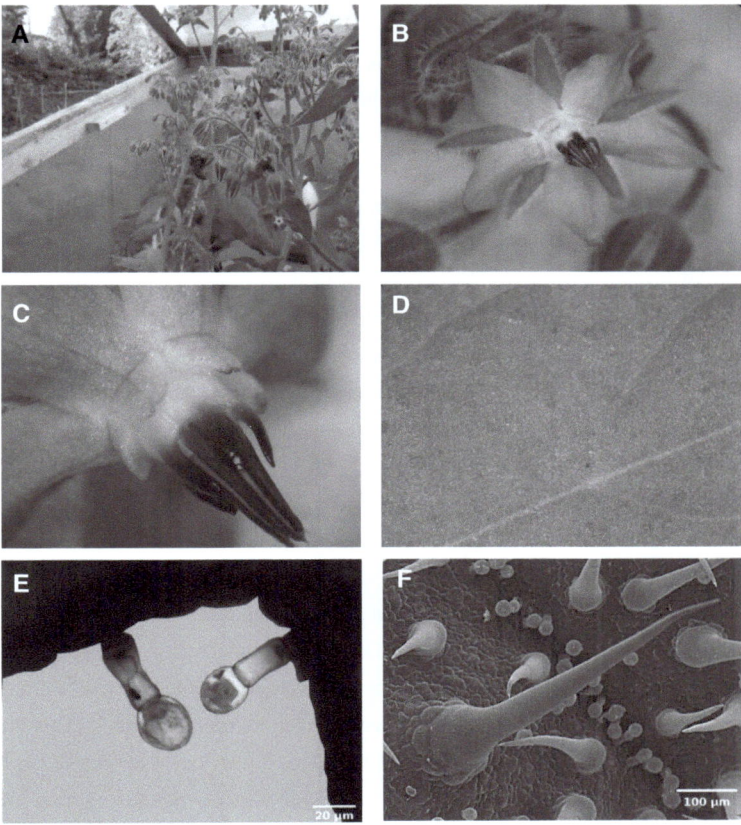

Figure 6. *Borago officinalis* L. (borage) **A)** Plant field-grown and flowering. Macroscopic image of **B)** a blue flower **C)** a lilac flower **D)** the upper leaf surface. **E)** Light microscopic images of glandular hairs (stained with FCA – fuchsin, chrysoidin and astrablue) (SCHÜLLER 2010) and **F)** SEM image of the leaf surface showing glandular and bristle hairs (SCHÜLLER 2010) (for magnitudes see scale bars).

In the oil of seeds more than 29 % gamma linolenic acid (GLA) was found, which makes borage also important for medical use (BARR 2001). For example, its leaves were reported to be diuretic, demulcent, emollient, expectorant, etc. which is partly due to the mucilage content (up to 30 %) (ESKIN & TAMIR 2006). However, the seeds of borage (*Borago officinalis* L.) are produced commercially as sources of GLA. 95% of the world's crop is grown in the United Kingdom, The Netherlands, Canada, New Zealand and Poland

(KAPOOR & NAIR 2005). Borage oil does not only contain GLA, but also erucic acid and toxic pyrrolizidine alkaloids (LARSEN & al. 1984, DESMET 1991). These pyrrolizidine alkaloids are extremely common in the Boraginaceae family and are powerful hepatotoxins. Although the total concentration in borage is extremely small (around 10 ppm in the dried herb), it has been argued that borage is an unsafe herb when used in folk medicine (KATZER 1999). Fortunately, the pyrrolizidine alkaloids are present at non-toxic levels (LARSEN & al. 1984, DESMET 1991) and therefore the risks associated with casual culinary usage are probably negligible (KATZER 1999). Antioxidant and ROS-scavenging properties of borage meal extract is attributed to the phenolic constituents whereas the dominant antioxidative compound was identified as rosmarinic acid (WETTASINGHE & al. 2001). The seed oil of borage is also containing linolic acid and γ-linolenic acid, saponins, mucilage, tannins, silicic acid and other essential oils and may act cardiotonic, expectorant and haemostatic (DACHLER & PELZMANN 1999, SULZBERGER 2002, MÜLLER 2004, VAUGHAN & JUDD 2006, MHAMDI & al. 2007).

The name borage can be traced back to Medieval Latin "borrago". The latter name is generally accepted to have Arabic origin – translated "father of sweat" or "father of roughness". In the first case, the motive would be the use of borage leaves in diaphoretic medicines, in the second case the rough leaf surface (*Fig. 6 D – F*).

Borage is a culinary herb mostly popular in Central Europe. The raw leaves are often used for salads or soups and also the blue flowers are edible and sometimes used dried as food colorant. Borage's taste is rather weak and very similar to fresh cucumber. This is also due to the fact that the dominating component in the essential oil of leaves is also a main component in cucumber aroma (cucumber aldehyde). In Germany in spring time often sauces are prepared from herbs. Most known, even outside Germany, is the *Green Sauce* made in Frankfurt which, beside borage, includes herbs like parsley, chervil, chives, cress, sorrel (*Rumex acetosa*) and burnet (KATZER 1999).

3.1.3 Summer savory

Summer savory (*Satureja hortensis* L.) is an annual, herbaceous aromatic medicinal plant belonging to the family of Lamiaceae (SEFIDKON & al. 2006), flowering from lilac to white (BICKERICH & al. 2001) and closely related to perennial winter savory (*Satureja montana* L.) (*Fig. 7 A – C*). Several species of the genus *Satureja* are found in the region around the Mediterranean Sea, although they probably originated from Western and Central Asia.

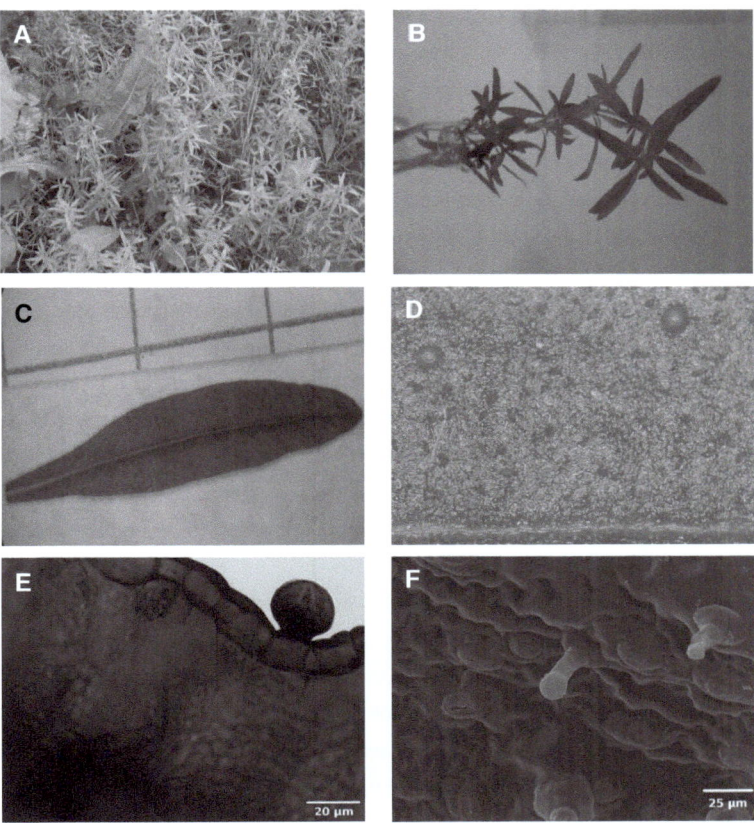

Figure 7. *Satureja hortensis* L. (summer savory) **A)** Plants field-grown **B)** Detail of the leafy stems **C)** detail of the leaf (millimeter paper in the background) **D)** macroscopic image of the leaf surface **E)** Light microscopic images of a plant cross section with glandular hair (stained with FCA – fuchsin, chrysoidin and astrablue) (SCHÜLLER 2010) and **F)** SEM image of the leaf surface showing glandular hairs and stomata (SCHÜLLER 2010) (for magnitudes see scale bars).

Savory is often used for bean dishes (therefore the German term "Bohnenkraut" literarily translated means "bean´s herb"). Several European languages name savory also as pepper herb, which is due to its former use as a substitute for black pepper (especially in Germany during World War II) (KATZER 2007). Even in ancient and medieval times savory was (probably) known as seasoning herb and the first mention of its essential oil was in 1582 (BICKERICH & al. 2001).

Summer savory's essential oils are including carvacrol, p-cymol, cymen, thymol, borneol, linalool, carvon and caryophyllen and can be found in glandular hairs on the leaf surface (BEZIĆ & al. 2009) (*Fig. 7 D – F*). Furthermore summer savory is containing tannins, bitter substances, resin, mucilage, potassium, calcium and sodium (DACHLER & PELZMANN 1999; HAJHASHEMI & al. 2000; BICKERICH & al. 2001; SULZBERGER 2002; MÜLLER 2004). BAHER & al. (2002) identified at least 17 components in the oils of *Satureja hortensis*, representing 99% of oil. Summer savory is also reported as source of rosmarinic acid, other phenols (ZHENG & WANG 2001), carnosol and carnosic acid (reviewed in YANISHLIEVA & al. 2006). It is used in traditional medicine (e.g. in Iran) for treating stomach and intestinal disorders (in BAHER & al. 2002) and especially the essential oils have demonst9rable antibacterial properties (BAHER & al. 2002, TAJKARIMI & al. 2010). Essential oils and extracts of *Satureja* species show (besides antibacterial) also antifungal properties and are therefore inhibiting human, food and plant pathogens (DEANS & SVOBODA 1989, HELANDER & al., 1998, HAJHASHEMI & al., 2000, GULLUCE & al. 2003, SAHIN & al. 2003, BAYDAR & al. 2004, BOYRAZ & ÖZCAN 2006). Even antiviral activity of savory's essential oils against HIV has been documented (YAMASAKI & al. 1998) and savory is supposed to be adipsous for diabedics (BICKERICH & al. 2001).

Savory is not only used as remedy but also consumed as vegetable (HAJHASHEMI & al. 2000) and is a popular seasoning herb. It is used both dried and fresh because of its spicy and a little bitter taste and is well known in the liqueur and perfume industry, too (DACHLER & PELZMANN 1999; HAJHASHEMI & al. 2000; BICKERICH & al. 2001; SULZBERGER 2002; MÜLLER 2004). The green leaves and herbaceous parts of stems of *Satureja hortensis* are used (fresh and dried) as flavouring agents in seasoning, stews, meat dishes, poultry, sausages and vegetables (SEFIDKON & al. 2004). The essential oils or the oleoresins are also used in baked goods, meat products, processed vegetables, condiment and relishes, soups, gravies and soft candy (BURDOCK 1995).

Material and Methods

3.2 Plant material

Three different plant species, which are mainly used as seasoning herbs, were used in experimentation:

Nasturtium *(Tropaeolum majus* L.), Tropaeolaceae, *Austrosaat*
Borage *(Borago officinalis* L.), Boraginaceae, *Austrosaat*
Summer savory (*Satureja hortensis* L.), Lamiaceae, *Austrosaat*

3.3 Meteorological data

Meteorological data from the meteorological station of the Karl-Franzens-University of Graz were edited from 1st of March to 31st of August 2008. Available data are described in *Tab. 1*.

Table 1: Data and units of meteorological parameters collected of the meteorological station of the Karl-Franzens-University of Graz between 1st of March and 31st of August 2008.

Data	factor	unit
Air temperature	arithmetic mean of 10 minutes	degree Celsius
Rain fall	summation of 10 minutes	milimeters
Relative humidity	arithmetic mean of 10 minutes	percentage
Air pressure	reduced to sea level	hPa
Sunshine duration	summation of 10 minutes	minutes
Sunrise/Sunset	time per day, UTC+1 and UTC+2	hours and minutes

Meteorological data (temperature, rainfall, humidity, air pressure and sunshine duration) were provided with kind permission of Mag. Dr. Erich PUTZ and Mag. Ing. Helga PIETSCH, times of sunrise and sunset of Univ.-Prof. Dr. Arnold HANSELMEIER (Geophysik, Astrophysik und Meteorologie (IGAM), Institut für Physik, Karl-Franzens-University of Graz).

3.4 Plant cultivation for experiments of drought stress

Thirty plants of nasturtium *(Tropaeolum majus* L.), summer savory (*Satureja hortensis* L.) and borage *(Borago officinalis* L.) were cultivated in climate chambers (seeds: *Austrosaat*, potting soil: *NaturaHum*, pot size: 9 x 9 x 9,5 cm, 3 – 10 plants per pot, once per week fertilized with *Wuxal* Top N, photoperiod: 12 h, PAR: 400–700 µmol m^{-2} s^{-1}, temperature: 22°C day/18°C night). After 6 weeks 10 plants were used as controls and 20 plants were drought stressed under defined conditions (30 % of the optimal water supply) for 7 days with a mild stress which assures that the defense network is activated but no morphological stress symptoms are shown. Thereafter drought stress was stopped by re-watering 10 of the pre-treated plants for 7 days. Selected plant material of controls, stressed and stressed/re-watered plants was harvested (summer savory – whole plants, borage and nasturtium – all leaves except tips), quick-frozen in liquid nitrogen, lyophilized and ground in a dismembrator. The resulting powder was subjected to the analyses described below.

3.5 Plant cultivation for field experiments

For drying, storage and processing experiments various plants of each species were sowed and field-grown under optimal conditions depending on plant species. Abiotic parameters were recorded by a meteorological station. Plant collections were conducted as shown in *Tab. 2*.

Table 2: Approaches and plant collections during the vegetation period in the field.

Date	Approach
27-03-2008	Sowing of nasturtium, borage, summer savory in field
24-06-2008	Measurement of photosynthetic parameters in field grown 3 month old flowering plants
01-07-2008/02-07-2008	Material harvests for drying incl. controls (flowering plants, 13 weeks old)
11-08-2008	Harvests for processing experiments (19 weeks old plants)

3.6 Drying and storage experiments

As it is common in the industrial processing plant material of each plant species was harvested before flowering at total differentiated leaves and dried with a drying hurdle at 35 °C for about one week depending on plant species (with kind permission of Dr. Hans BERGHOLD, Institute of Sustainable Techniques and Systems, Chemical and Technical Plant Utilization, Joanneum Research Forschungsgesellschaft mbH).

The plant material was stored in two different ways: dried in paper bags and dried in tie bags (paper bags equipped with foil - as it is used for storage of teas) in the dark with constant room temperature. Ingredients were examined 3 months after storage in borage and summer savory and 6 months after storage in nasturtium (*Tab. 3*).

Table 3: Names, abbreviations and descriptions of the storage experiments of nasturtium.

Name (abbreviation)	Storage description
Fresh controls (CF)	fresh field grown plant material was shock frozen
Dried control (CD)	Hurdle dried plant material was shock frozen
3 or 6 months tie bag (3/6MT)	Hurdle dried plant material was stored for 3 or 6 months in tie bags (equipped with foil) before shock freezing
3or 6 months paper bag (3/6MP)	Hurdle dried plant material was stored for 3 or 6 months in paper bags before shock freezing

3.7 Processing experiments

For processing experiments dried plant material as well as fresh leaves were used. Fresh material was harvested directly before flowering of each plant species as it is common in the industrial processing. Processing was done as described in *Tab. 4* below.

Table 4: Names, abbreviations and descriptions of the processing experiments of nasturtium.

Name (abbreviation)	Processing description
Fresh controls (CF)	fresh field grown plant material was shock frozen
Cut (C)	fresh field grown plant material was cut and immediately shock frozen
Cut and 10 min resting (CR)	fresh field grown plant material was cut and let rested for 10 minutes before shock freezing
Cut and 5 min boiled (C5)	fresh field grown plant material was cut and 5 minutes boiled before shock freezing
Cut and 20 min boiled (C20)	fresh field grown plant material was cut and 20 minutes boiled before shock freezing
Cut, rested and 5 min boiled (CR5)	fresh field grown plant material was cut, let rested for 10 minutes and afterwards boiled for 5 minutes before shock freezing
Cut, rested and 20 min boiled (CR20)	fresh field grown plant material was cut, let rested for 10 minutes and afterwards boiled for 20 minutes before shock freezing
Dried control (CD)	Hurdle dried plant material
Dried and 5 min boiled (DB5)	dried plant material was 5 minutes boiled before shock freezing
Dried and 20 min boiled (DB20)	dried plant material was 20 minutes boiled before shock freezing

3.8 Biochemical analyses

For biochemical studies plant material was put into paper bags, quick-frozen in liquid nitrogen, lyophilized and ground in a dismembrator. The resulting powder was subjected to the analyses described below.

3.8.1 Determination of plastid pigments (modified according to PFEIFHOFER 1989)

All important chloroplast pigments (neoxanthin, violaxanthin, antheraxanthin, lutein, zeaxanthin, chlorophyll a, chlorophyll b, α-carotene, β-carotene) were separated and determined in one step using the HPLC (high-performance liquid chromatography) gradient method. Pulverized plant-material (50-60 mg) was added to 60 mg calcium carbonate and extracted with DMSO/ethanol (Dimethylsulfoxide,2:1, v:v) 3 times on ice in the dark. Before the transfer into the HPLC vials the samples were centrifuged for 30 minutes at 4°C at 14,000 rpm. Separation and determination of the pigments was done on a gradient HPLC (HP Chemstation, 4°C cooled autosampler, used column 25 x 4.6 mm Grom Spherisorb ODS2 5 μm, photometric detection by HP diode array detector 1040 M at 440 nm). Solvent A: acetonitrile/aqua bidest./methanol (100/10/5, v/v/v); Solvent B: acetone/ethyl acetate (2/1, v/v); Gradient: 10 % to 80 % solvent B in 18 seconds. Flow rate: 1 ml min^{-1}.

3.8.2 Determination of tocopherol (vitamin E) (modified according to WILDI & LÜTZ 1996)

For the determination of tocopherol the same extraction procedure as for the pigment analyses was used (modified according to PFEIFHOFER 1989). HPLC was done using methanol as the solvent. Software ChromStar 4.0; Column: 25 x 4.6 mm Grom Spherisorb ODS2 5 μm, fluorescence-detector *Jasco* FP 2020 Plus, emission 325 nm, excitation 295 nm wavelength, 4 °C autosampler; flow rate 1 ml min^{-1}.

3.8.3 Determination of ascorbic acid (modified according to TAUSZ & al. 1996)

For determination of ascorbic acid the total ascorbic acid content (reduction of the oxidized ascorbic acid with DTT) as well as the content of reduced ascorbic acid was determined by high performance liquid chromatography (HPLC). Thereafter the content of oxidized ascorbic acid was calculated.

For extraction meta-phosphoric acid 1.5 % was added to PVPP (polyvinylpolypyrrolidone) and incubated for a few hours. 60 – 80 mg powdered plant material was added to this solution. This extraction was used for two different approaches (approach 1 and 2). For determination of reduced ascorbic acid (approach 1), TRIS-buffer (0.2 M Tris(hydroxymethyl)aminomethan and 1 mM EDTA in aqua bidest.), aqua bidest. and ortho-phoshoric-acid (85% ortho-phosphoric acid/aqua bidest, 1:10, v:v) were added to the extract. For determination of total ascorbic acid content (approach 2) TRIS-buffer and DTT-solvent (60 mg Dithiothreitol in 1.5 ml aqua bidest.) were added to the extract. This solution (approach 2) was incubated for 10 minutes at room temperature. Reactions were stopped by addition of ortho-phosphoric-acid. After centrifugation the supernatants were transferred into HPLC vials. HPLC analyses were done with a solvent of 50 mM $NH_4H_2PO_4$/acetonitrile (30/70, v/v). Software ChromStar 4.0; Column: GROM-SIL 120 Amino-2PA, 5 µm, 250 x 4.6 mm, photometric detection by an UV/Vis detector from *SunChrom*, SpectraFlow 505 at a wavelength of 250 nm, 4 °C cooled autosampler, flow rate 1 ml min^{-1}.

3.8.4 Determination of oxidized and reduced glutathione and cysteine (modified to KRANNER & GRILL 1993)

Total contents of thiols were analyzed by a modified HPLC method according to KRANNER & GRILL (1993).

60-80 mg lyophilized powder of plant material was extracted in 3 ml 0.1 M HCl prepared with 0.06 g polyvinylpolypyrrolidone (PVPP), which was swelled in the HCl-solution at least for 12 hours. The extract was homogenized with an Ultraturax (20sec., 24,000 rpm) and cooled centrifuged for 10 minutes at 14,000 rpm.

Total thiols: 280 µl supernatant of the centrifuged extract were added to 30 µl of sodium hydroxid solution (1M), 280 µl of tricine buffer (200mM; 1,79 g tricine and 20 mg EDTA in 50 ml aqua bidest, pH 8.0) and 70 µl of DTT (10 mM dithiothreitol in aqua bidest.) in a dark reaction tube, vortexed (pH of the mixture between 7.9 and 8.3) and incubated for one hour at room temperature. The SH-groups of the reduced extract were marked with 100 µl monobromobimane (8 mM in acetonitril and incubated for 15 minutes. Afterwards the reaction was stopped by addition of 600 µl MSA (0.75 % methansulfonic acid in water). The derivated extract was centrifuged for 30 minutes at 4°C and 14,000 rpm and subjected to the HPLC analysis.

Oxidized thiols: For the determination of the content of oxidized thiols 400 µl supernatant of the centrifuged extract was added to 43 µl of sodium hydroxide solution (1M), 400 µl of tricine buffer (200 mM, pH 8.0) and 30 µl NEM (N-ethylmaleimide, 50 mM in isopropanol) in a transparent reaction tube, mixed and subsequently incubated at room temperature for

15 minutes in order to block the SH-groups. The excess of NEM was removed by extraction with 500 µl toluene (3 times, centrifugation for 3 minutes). After addition of 57 µl DTT to 500 µl of the new extract the mixture was incubated in the dark for one hour at room temperature. The now reduced SH-groups were labeled with 82 µl monobromobimane (8 mM in acetonitrile) and incubated for 15 minutes in the dark at room temperature. The derivatization was stopped by addition of 491 µl MSA. For the subjection to the HPLC the extracts were centrifuged for 30 minutes at 4°C and 14,000 rpm.

Separation and determination of the derivatized thiols were conducted on a gradient HPLC-systems: *Chromstar* Software 4.0, *Knauer* Degasser, two *SunChrom* HPLC-pumps SunFlow 100, *Knauer* mixer, *Spark Holland Midias* autosampler cooled at 4°C, *Jasco* Flourescence dedector FP 2020 plus (excitation 380 nm, emission 480 nm), Grom Spherisorb ODS-2 250 x 4.6 mm 50µm column with Grom Spherisorb ODS-2 5 x 4.6 mm pre-column. Gradient: solvent A: aqua bidest./methanol/acetic acid (1000/50/2.5; v/v/v), pH 3.9), solvent B: aqua bidest./methanol (1/9; v/v). 7 % solvent B to 20 % solvent B in 20 minutes, 20 % solvent B to 100 % solvent B in 0.5 minutes, 100 % solvent B for 6.5 minutes, 100 % solvent B to 7 % solvent B in 0.5 minutes, and 7 % solvent B for another 7.5 minutes, flow rate: 1 ml min^{-1}, Injection volume 50 µl.

3.9 Determination of photosynthetic parameters using the LI-6400 (LI-COR 1998)

Photosynthetic parameters (net-photosynthesis, transpiration, stomatal conductance, intercellular CO_2 concentration, light response and CO_2 response curves) were conducted with an IRGA (Infra-Red Gas Analyzer), the LI-6400. It is an open system, which means that measurements of photosynthesis and transpiration are based on the differences in CO_2 and H_2O in an air stream that is flowing through the leaf cuvette. Photosynthesis and transpiration were computed from the differences in CO_2 and H_2O between in-chamber conditions and pre-chamber conditions. Data for light and CO_2 response curves as well as net-photosynthesis, transpiration, stomatal conductance and C_i (intercellular CO_2 concentration) were prepared according to LI-COR manual (1998) and HOPFER (2007).

4. Drought stress experiments

4.1 Results of nasturtium

Both, contents of various substances and photosynthetic parameters were examined in control plants, drought stressed plants and re-watered plants.

4.1.1 Pigments

Total content of chlorophyll a was between 8,000 and 11,000 µg / g dry weight and significantly higher in stressed plants than in controls or re-watered plants. Between controls and re-watered plants no significant differences were found. Total content of chlorophyll b was between 2,000 and 3,000 µg/g dry weight and significantly higher in stressed plants compared to both, controls and re-watered plants, significantly lowest content of chlorophyll b was found in the re-watered plants (*Fig. 8*).

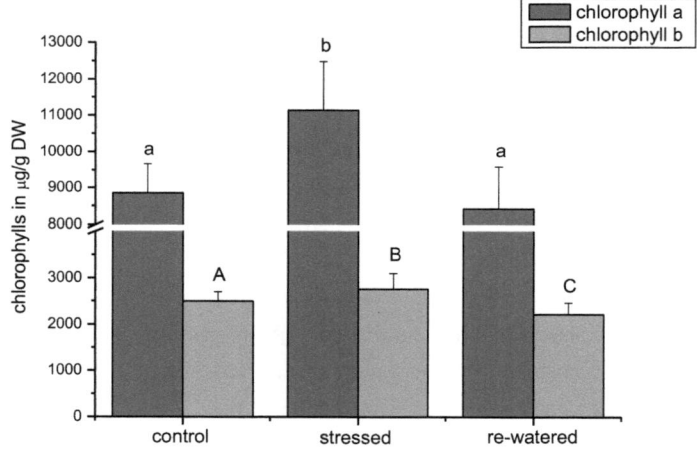

Figure 8: Total content of chlorophyll a and b in control plants, stressed and re-watered plants. Significant differences between the samples were indicated by different letters. P < 0.05 analyzed by Kruskal-Wallis-ANOVA, n = 10, error bars show standard deviation, DW = dry weight.

Content of alpha-carotene was about 20 and 40 µg / g dry weight and there was no significant difference between controls, stressed and re-watered plants. Beta-carotene values were between 700 and 1,000 µg / g dry weight and significantly higher in stressed plants compared to both, control and re-watered plants (*Fig. 9*).

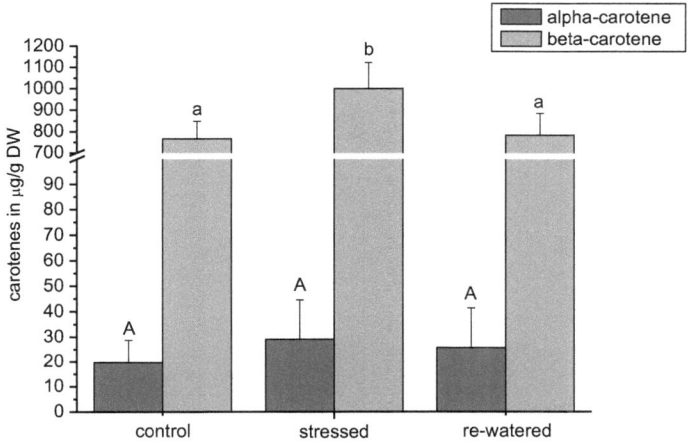

Figure 9: Total content of alpha- and beta-carotene in control plants, stressed and re-watered plants. Significant differences between the samples were indicated by different letters, same letters mean no significant differences. P < 0.05 analyzed by Kruskal-Wallis-ANOVA, n = 10, error bars show standard deviation, DW = dry weight.

Figure 10 shows that there were no significant differences in the total violaxanthin content of all three variants (all contents about 150 µg / g dry weight) but neo-, anthera- and lutein+zeaxanthin contents were significantly higher in stressed nasturtium plants compared to both, controls and re-watered plants. Antheraxanthin in re-watered plants was significantly lower than in stressed plants but also significantly higher than in the examined controls.

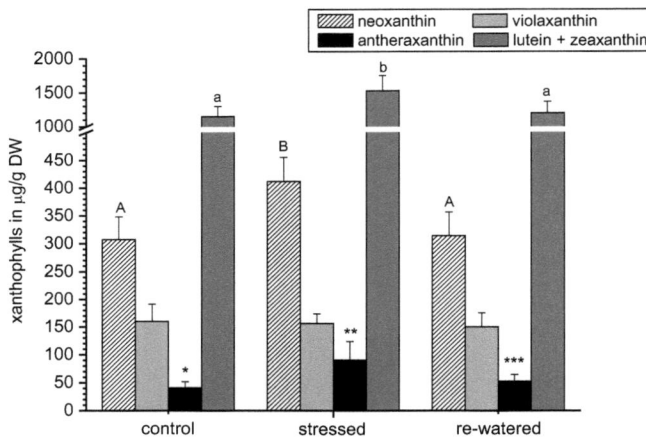

Figure 10: Total content of xanthophylls (neoxanthin, violaxanthin, antheraxanthin and lutein+zeaxanthin) in controls, stressed and re-watered plants. Significant differences between the samples were indicated by different letters and signs, no letters or signs mean no significant differences. P < 0.05 analyzed by Kruskal-Wallis-ANOVA, n = 10, error bars show standard deviation, DW = dry weight.

The ratio of chlorophyll a / b was between 3.5 and 4.5 and significantly higher in stressed than control plants. There was no significant difference between controls and re-watered plants as well as between stressed and re-watered plants (*Fig. 11*).

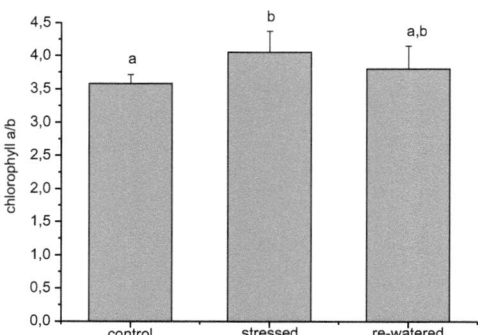

Figure 11: Ratio of chlorophyll a / b for controls, stressed and re-watered plants. Significant differences between the samples were indicated by different letters. P < 0.05 analyzed by Kruskal-Wallis-ANOVA, n = 10, error bars show standard deviation, DW = dry weight.

4.1.2 Tocopherol

Leaves of nasturtium plants showed significant differences in α-tocopherol in controls, stressed and re-watered plants. More α-tocopherol was found in stressed plants than in controls and above all the re-watered plants showed twice higher α-tocopherol contents than the control plants (*Fig. 12*).

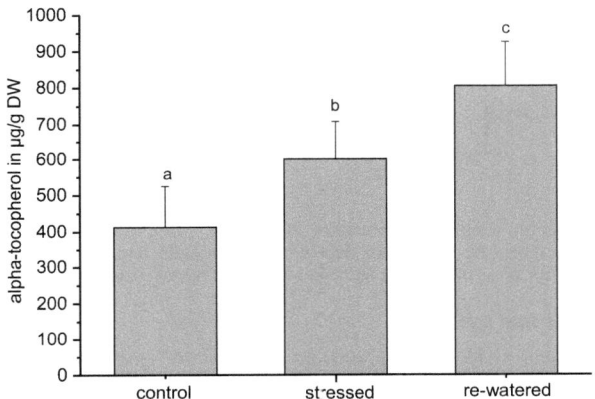

Figure 12: Total content of alpha-tocopherol in controls, stressed and re-watered plants. Significant differences between the samples were indicated by different letters. $P < 0.05$ analyzed by Kruskal-Wallis-ANOVA, n = 10, error bars show standard deviation, DW = dry weight.

4.1.3 Ascorbate

As *Figure 13* shows, there were no significant differences according to the ascorbate contents between all three variants. Content of total ascorbate was about 7,000 to 8,000 µg / g dry weight and not changing during stress or recovery.

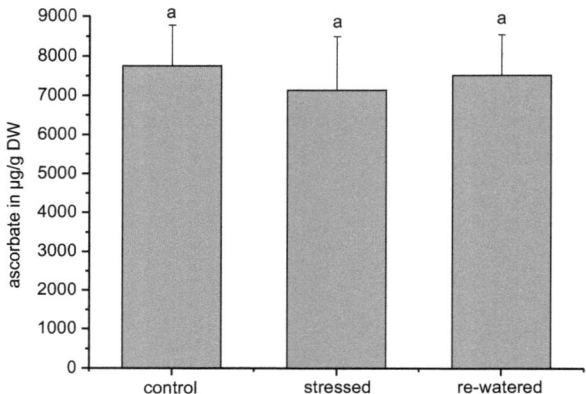

Figure 13: Total content of ascorbate in controls, stressed and re-watered plants. Same letters mean no significant differences. P < 0.05 analyzed by Kruskal-Wallis-ANOVA, n = 10, error bars show standard deviation, DW = dry weight.

4.1.4 Glutathione and cysteine

In nasturtium leaves cysteine contents were between 200 and 300 nmol / g dry weight and significantly higher in re-watered plants compared to controls. There was no significant difference found in controls compared to stressed plants or stressed to re-watered plants (*Fig. 14*).

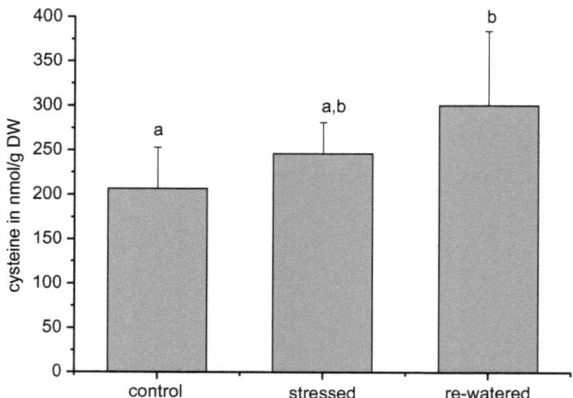

Figure 14: Total content of cysteine in controls, stressed and re-watered plants. Significant differences between the samples were indicated by different letters. P < 0.05 analyzed by Kruskal-Wallis-ANOVA, n = 10, error bars show standard deviation, DW = dry weight.

Figure 15 shows the proportional distribution of oxidized and reduced cysteine in nasturtium leaves of controls, stressed and re-watered plants calculated relating to total cysteine amounts. Oxidized cysteine was at about 75 % in controls and significantly decreasing during stress to about 55 % and not significantly increasing during re-watering up to about 75 %.

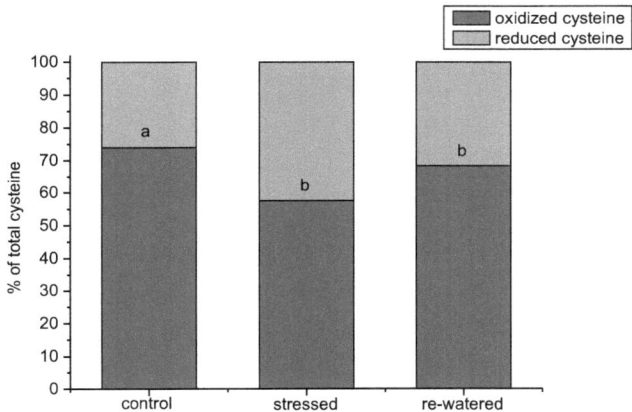

Figure 15: Percentage of oxidized and reduced cysteine in controls, stressed and re-watered plants. Percentage was calculated relating to total cysteine amounts. Significant differences between the samples were indicated by different letters. $P < 0.05$ analyzed by Kruskal-Wallis-ANOVA, n = 10.

Contents of total glutathione were between 1,500 and 3,500 nmol / g dry weight and significantly higher in stressed plants compared to controls and re-watered plants. In re-watered plants total glutathione contents were significantly lower than in controls (*Fig.16*).

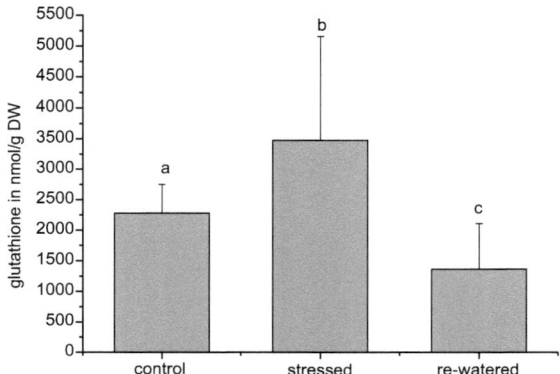

Figure 16: Total content of glutathione in controls, stressed and re-watered plants. Significant differences between the samples were indicated by different letters. P < 0.05 analyzed by Kruskal-Wallis-ANOVA, n = 10, error bars show standard deviation, DW = dry weight.

Figure 17 shows the proportional distribution of oxidized and reduced glutathione in nasturtium leaves of controls, stressed and re-watered plants calculated relating to total glutathione amounts. Oxidized glutathione was about 55 % in controls and significantly decreasing during stress to about 45 % and increasing again during re-watering up to 75 % and was therefore significantly higher than in controls.

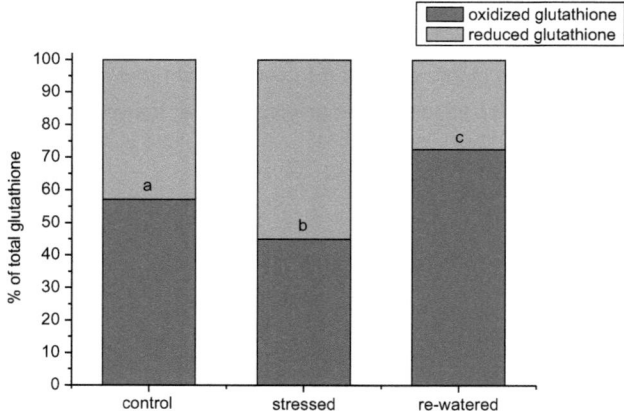

Figure 17: Percentage of oxidized and reduced glutathione in controls, stressed and re-watered plants. Percentage was calculated relating to total glutathione amounts. Significant differences between the samples were indicated by different letters. P < 0.05 analyzed by Kruskal-Wallis-ANOVA, n = 10.

4.1.5 Photosynthetic parameters

Photosynthetic parameters were illustrated in *Fig. 18*. Intercellular CO_2 concentration (C_i) was in controls about 250 µmol CO_2 mol^{-1} and significantly higher in stressed plants (about 350 µmol CO_2 mol^{-1}) compared to both, controls and re-watered plants and lowest in re-watered plants (about 150 µmol CO_2 mol^{-1}). Assimilation was about 7 µmol CO_2 m^{-2} s^{-1} and not significantly different in controls and re-watered plants but significantly lower in stressed plants (about 2 µmol CO_2 m^{-2} s^{-1}). Transpiration was highest in controls (about 1.2 mmol H_2O m^{-2} s^{-1}), decreasing to about 0.5 mmol H_2O m^{-2} s^{-1} during drought stress and slightly increasing during re-watering up to about 0.7 mmol H_2O m^{-2} s^{-1} and was therefore significantly higher than in stressed but lower than in control plants. Stomatal conductance was significantly highest in controls (about 0.04 mol H_2O m^{-2} s^{-1}), decreasing during stress to about 0.015 mol H_2O m^{-2} s^{-1} and slightly increasing during recovery up to about 0.025 mol H_2O m^{-2} s^{-1} and was therefore significantly lowest in stressed plants.

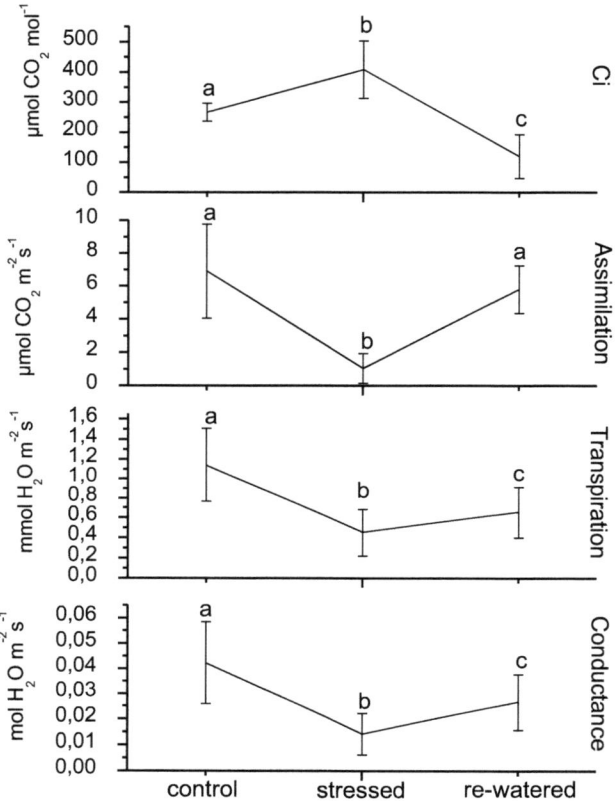

Figure 18: Ci, assimilation, transpiration and stomatal conductance of controls, stressed and re-watered nasturtium plants. Significant differences between the samples were indicated by different letters. $P < 0.05$ analyzed by Kruskal-Wallis-ANOVA, n = 10, error bars show standard deviation. Ci = intercellular CO_2 concentration.

Figure 19 shows the light curves of controls, stressed and re-watered nasturtium plants. Assimilation rates of stressed plants were hardly changing during increasing light intensities and are mostly below zero. The light compensation point in controls was at about 50 µmol photons m^{-2} s^{-1}, in re-watered plants at about 20 µmol photons m^{-2} s^{-1}. Controls showed slowly but steadily increasing assimilation rates and re-watered plants fast increasing and highest assimilation rates which were slowly decreasing to control levels.

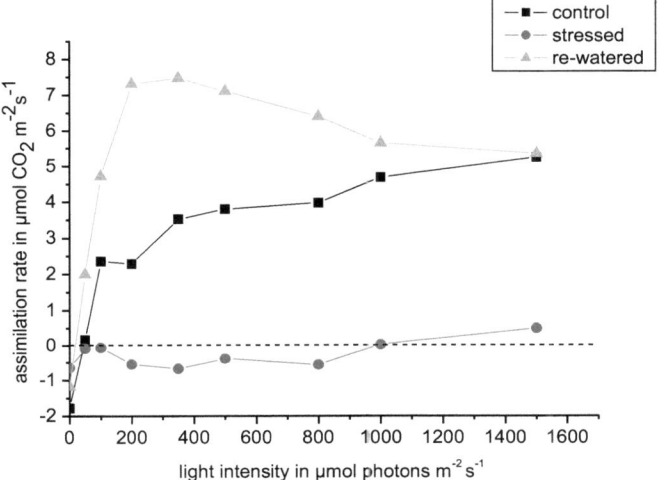

Figure 19: Light curve – Changes of assimilation rates during increasing light intensity in controls, stressed and re-watered nasturtium plants. Dotted line shows the light compensation points.

Transpiration rate and stomatal conductance of stressed plants were hardly changing during increasing light intensities. Control plants showed slow but steady increase, re-watered plants fast increase and at about 400 PAR slow decrease but all in all higher transpiration rates and stomatal conductance compared to both, controls and stressed plants (*Fig.20 A and B*).

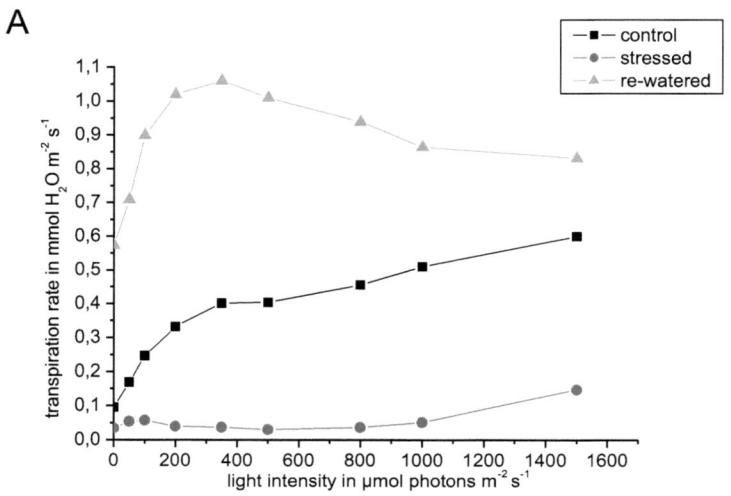

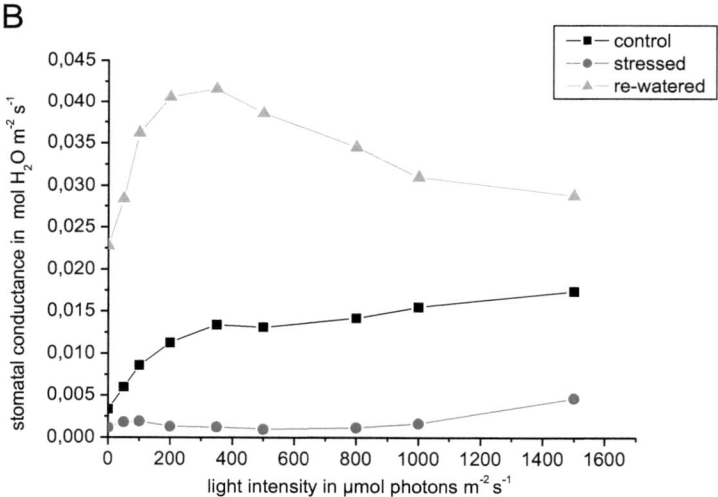

Figure 20: Changes of transpiration rates (A) and stomatal conductance (B) during increasing light intensity in controls, stressed and re-watered nasturtium plants.

4.1.6 Summary of the results of nasturtium

Table 5 shows the significant changes of all examined substances as an overview and summary.

Table 5: Summary of significant changes in control, stressed and re-watered nasturtium plants. ↑ = increase, ↓ = decrease, ↔ = no significant change, black arrows: compared to control plants, red arrows in brackets: compared to stressed plants.

Nasturtium	stressed	re-watered	
chlorophyll a	↑	↔	(↓)
chlorophyll b	↑	↓	(↓)
α-carotene	↔	↔	(↔)
β-carotene	↑	↔	(↓)
neoxanthin	↑	↔	(↓)
violaxanthin	↔	↔	(↔)
antheraxanthin	↑	↑	(↓)
lutein+zeaxanthin	↑	↔	(↓)
chlorophyll a/b	↑	↔	(↔)
α-tocopherol	↑	↑	(↑)
ascorbate	↔	↔	(↔)
cysteine	↔	↑	(↔)
glutathione	↑	↓	(↓)
Ci	↑	↓	(↓)
assimilation	↓	↔	(↑)
transpiration	↓	↓	(↑)
stomatal conductance	↓	↓	(↑)

4.2 Results of summer savory

Contents of various substances were examined in control plants, drought stressed plants and re-watered plants.

4.2.1 Pigments

Total content of chlorophyll a was between 4,000 and 6,000 µg / g dry weight and about equal in stressed and control plants but significantly lower in re-watered plants. Total content of chlorophyll b was at about 1,500 µg/g dry weight and significantly lower in re-watered plants compared to both, controls and stressed plants. No significant differences were found in controls compared to stressed plants (*Fig. 21*).

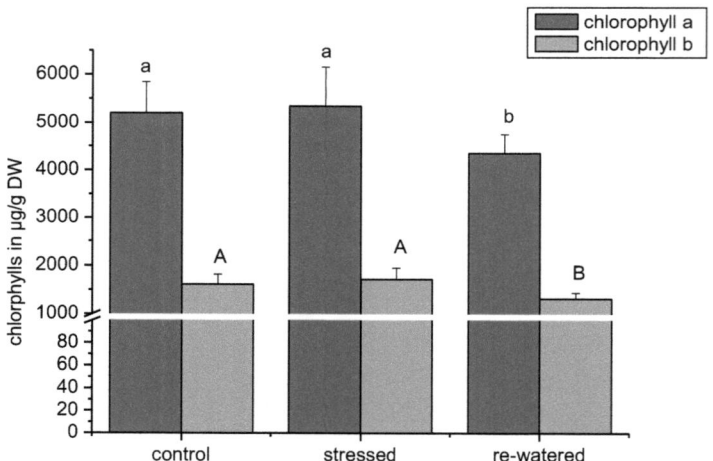

Figure 21: Total content of chlorophyll a and b in control plants, stressed and re-watered plants. Significant differences between the samples were indicated by different letters. $P < 0.05$ analyzed by Kruskal-Wallis-ANOVA, n = 10, error bars show standard deviation, DW = dry weight.

Content of alpha-carotene was about 3 and 10 µg / g dry weight and there was no significant difference between controls and stressed plants as well as between stressed and re-watered plants but controls showed significant higher contents compared to re-watered plants. Beta-carotene values were between 300 and 400 µg / g dry weight and significantly higher in controls and stressed plants compared to re-watered plants. No significant difference in beta-carotene content was found between stressed and control plants (*Fig. 22*).

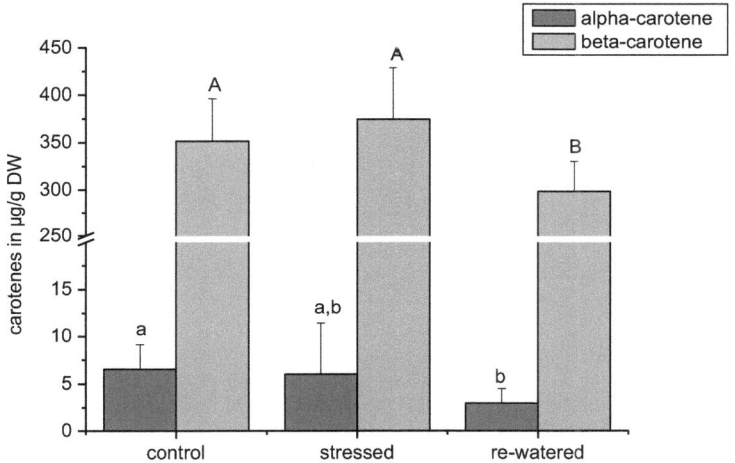

Figure 22: Total content of alpha- and beta-carotene in control plants, stressed and re-watered plants. Significant differences between the samples were indicated by different letters, same letters mean no significant differences. $P < 0.05$ analyzed by Kruskal-Wallis-ANOVA, n = 10, error bars show standard deviation, DW = dry weight.

Figure 23 shows that there were no significant differences in the total antheraxanthin content of all three variants (all contents about 50 µg / g dry weight) but neo-, and lutein+zeaxanthin contents were significantly lower in re-watered nasturtium plants compared to both, controls and stressed plants. Violaxanthin was significantly higher in controls than in stressed or re-watered plants.

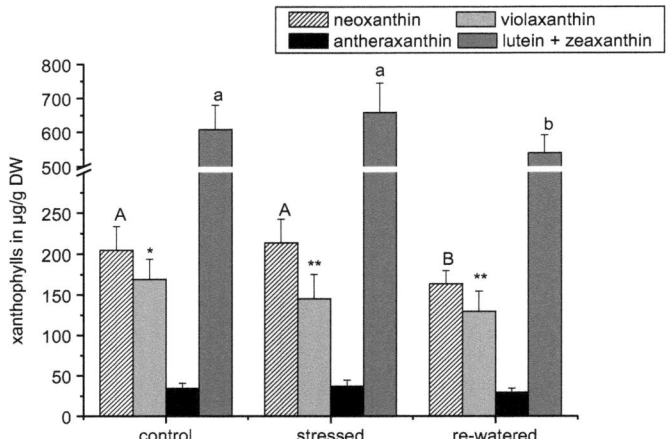

Figure 23: Total content of xanthophylls (neoxanthin, violaxanthin, antheraxanthin and lutein+zeaxanthin) in controls, stressed and re-watered plants. Significant differences between the samples were indicated by different letters and signs, no letters or signs mean no significant differences. P < 0.05 analyzed by Kruskal-Wallis-ANOVA, n = 10, error bars show standard deviation, DW = dry weight.

The ratio of chlorophyll a / b was between 3.0 and 3.5 there was no significant difference between controls, stressed and re-watered plants (*Fig. 24*).

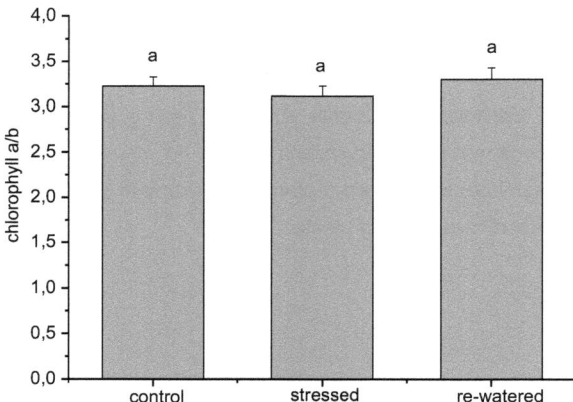

Figure 24: Ratio of chlorophyll a/b for controls, stressed and re-watered plants. Significant differences between the samples were indicated by different letters. P < 0.05 analyzed by Kruskal-Wallis-ANOVA, n = 10, error bars show standard deviation, DW = dry weight.

4.2.2 Tocopherol

Leaves of summer savory plants showed significant different high contents of α-tocopherol in controls and re-watered plants but not in controls compared to stressed as well as stressed compared to re-watered plants. All contents were between 300 and 400 µg / g dry weight (*Fig. 25*).

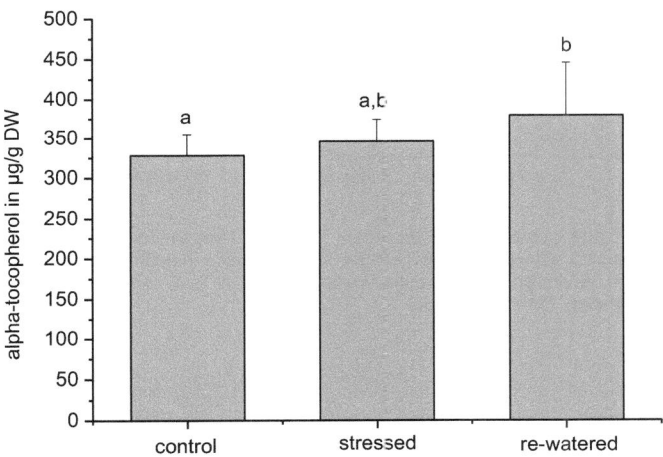

Figure 25: Total content of alpha-tocopherol in controls, stressed and re-watered plants. Significant differences between the samples were indicated by different letters. $P < 0.05$ analyzed by Kruskal-Wallis-ANOVA, n = 10, error bars show standard deviation, DW = dry weight.

4.2.3 Ascorbate

As *Fig. 26* shows, there was no significant difference according to the ascorbate contents of stressed compared to re-watered plants but in both contents were significantly higher than in controls. Content of total ascorbate was about 1,700 to 2,700 µg / g dry weight.

The proportional distribution of oxidized and reduced ascorbate in summer savory leaves of controls, stressed and re-watered plants was calculated relating to total ascorbate amounts and shown in *Fig. 27*. Reduced ascorbate was about 60 % in controls and stressed plants but significantly increasing during re-watering to about 85 %.

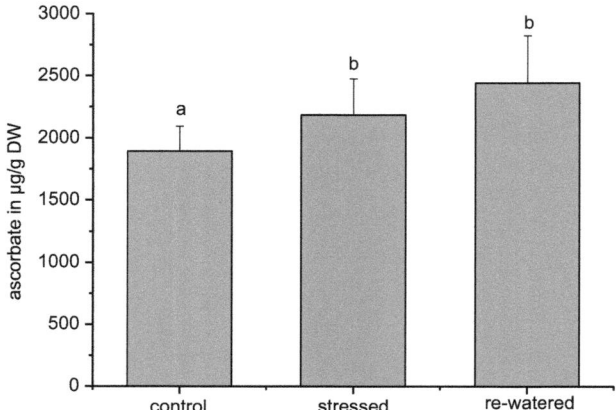

Figure 26: Total content of ascorbate in controls, stressed and re-watered plants. Significant differences between the samples were indicated by different letters. P < 0.05 analyzed by Kruskal-Wallis-ANOVA, n = 10, error bars show standard deviation, DW = dry weight.

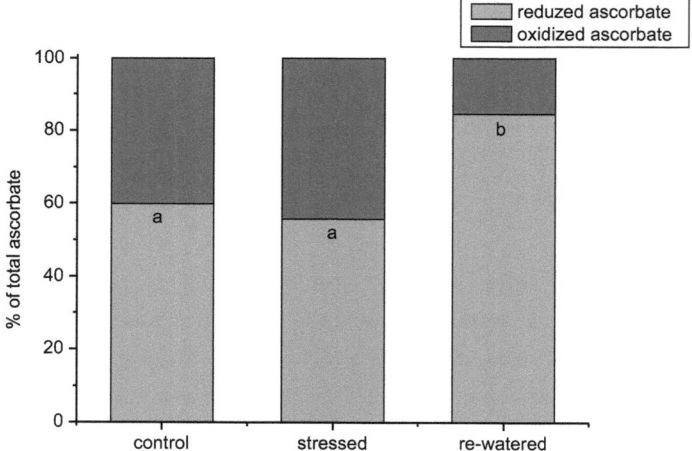

Figure 27: Percentage of oxidized and reduced ascorbate in controls, stressed and re-watered plants. Percentage is calculated relating to total ascorbate amounts. Significant differences between the samples were indicated by different letters. P < 0.05 analyzed by Kruskal-Wallis-ANOVA, n = 10.

4.2.4 Glutathione

Contents of total glutathione were between 2,500 and 5,000 nmol / g dry weight and significantly lower in re-watered compared to stressed plants. No significant differences were found between controls and both stressed and re-watered plants (*Fig. 28*).

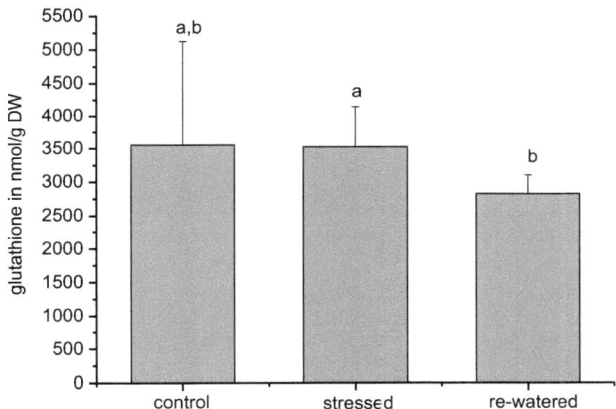

Figure 28: Total content of glutathione in controls, stressed and re-watered plants. Significant differences between the samples were indicated by different letters. $P < 0.05$ analyzed by Kruskal-Wallis-ANOVA, n = 10, error bars show standard deviation, DW = dry weight.

Figure 29 shows the proportional distribution of oxidized and reduced glutathione in controls, stressed and re-watered plants calculated relating to total glutathione amounts. Oxidized glutathione was about 40 % in controls and stressed plants and significantly different compared to re-watered plant. During stress and recovery the percentage of oxidized glutathione was decreasing.

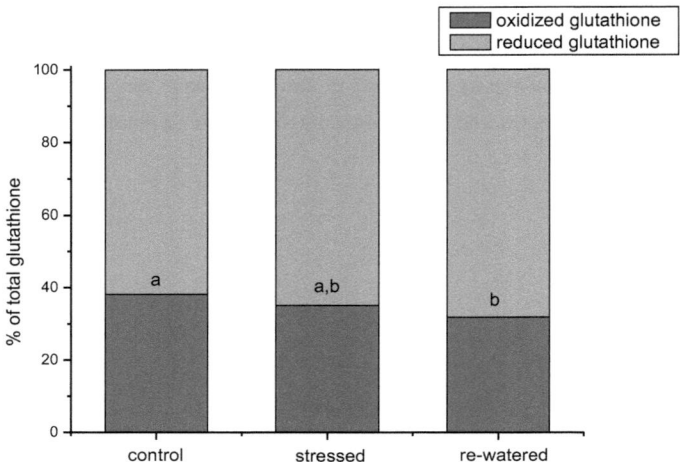

Figure 29: Percentage of oxidized and reduced glutathione in controls, stressed and re-watered plants. Percentage is calculated relating to total glutathione amounts. Significant differences between the samples were indicated by different letters. P < 0.05 analyzed by Kruskal-Wallis-ANOVA, n = 10.

4.2.5 Summary of the results of summer savory

Table 6 shows the significant changes of all examined substances as an overview and summary.

Table 6: Summary of significant changes in control, stressed and re-watered summer savory plants. ↑ = increase, ↓ = decrease, ↔ = no significant change, black arrows: compared to control plants, red arrows in brackets: compared to stressed plants.

Summer savory	stressed	re-watered	
chlorophyll a	↔	↓	(↓)
chlorophyll b	↔	↓	(↓)
α-carotene	↔	↓	(↓)
β-carotene	↔	↓	(↓)
neoxanthin	↔	↓	(↓)
violaxanthin	↓	↓	(↔)
antheraxanthin	↔	↔	(↔)
lutein+zeaxanthin	↔	↓	(↓)
chlorophyll a/b	↔	↔	(↔)
α-tocopherol	↔	↑	(↔)
ascorbate	↑	↑	(↔)
glutathione	↔	↓	(↑)

4.3 Results of borage

Contents of various substances were examined in control plants, drought stressed plants and stressed and re-watered plants.

4.3.1 Pigments

Total content of chlorophyll a was between 4,500 and 5,500 µg / g dry weight and not significantly different in all three variants. Total content of chlorophyll b was at about 1,200 µg/g dry weight, no significant differences were found (*Fig. 30*).

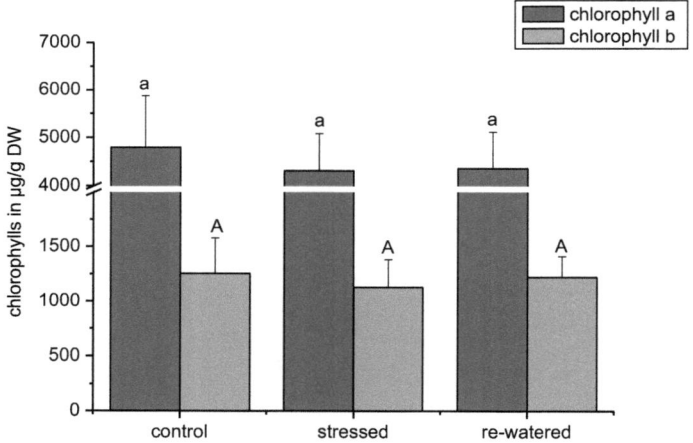

Figure 30: Total content of chlorophyll a and b in control plants, stressed and re-watered plants. Same letters mean no significant difference. P < 0.05 analyzed by Kruskal-Wallis-ANOVA, n = 10, error bars show standard deviation, DW = dry weight.

Content of alpha-carotene was about 2 and 6 µg / g dry weight and there was no significant difference between controls and stressed plants but contents in re-watered plants were significantly increasing compared to both. Beta-carotene values were between 400 and 600 µg / g dry weight but showed no significant difference in all three variants (*Fig. 31*).

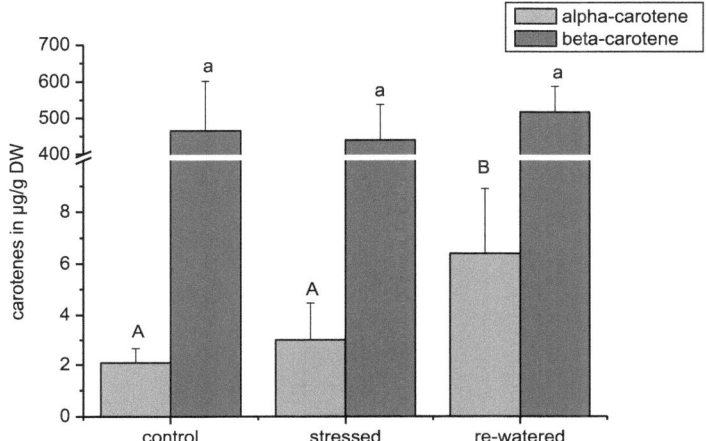

Figure 31: Total content of alpha- and beta-carotene in control plants, stressed and re-watered plants. Significant differences between the samples are indicated by different letters, same letters mean no significant differences. P < 0.05 analyzed by Kruskal-Wallis-ANOVA, n = 10, errors bars show standard deviation, DW = dry weight.

Figure 32 shows that there were no significant differences in the total neo-, viola-, anthera- and lutein+zeaxanthin contents of all three variants. Neoxanthin contents were about 200 µg, violaxanthin 300 µg, antheraxanthin 25 µg and lutein+zeaxanthin about 600 µg / g dry weight.

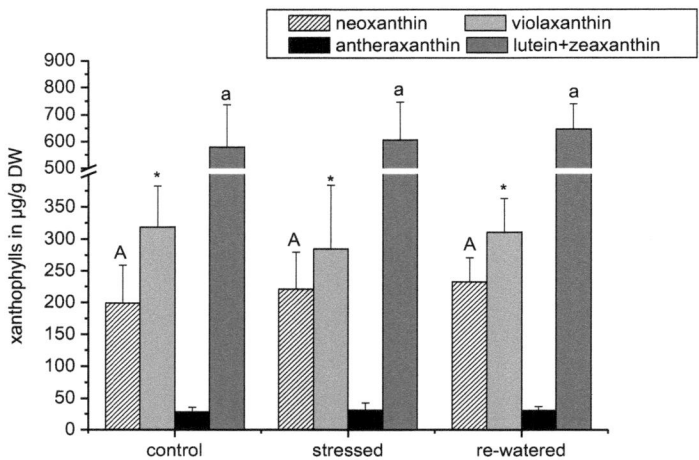

Figure 32: Total content of xanthophylls (neoxanthin, violaxanthin, antheraxanthin and lutein+zeaxanthin) in controls, stressed and re-watered plants. Same or no letters and signs mean no significant difference. P < 0.05 analyzed by Kruskal-Wallis-ANOVA, n = 10, error bars show standard deviation, DW = dry weight.

The ratio of chlorophyll a / b was between 3.5 and 4.5 but not significantly different in all three variants (*Fig. 33*).

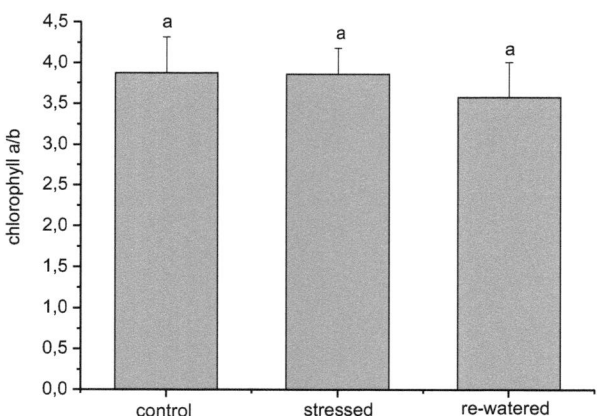

Figure 33: Ratio of chlorophyll a / b for controls, stressed and re-watered plants. Same letters mean no significant differences. P < 0.05 analyzed by Kruskal-Wallis-ANOVA, n = 10, error bars show standard deviation, DW = dry weight.

4.3.2 Tocopherol

Leaves of borage plants showed significantly higher contents of α-tocopherol in controls and both, stressed and re-watered plants. All contents were between 200 and 350 µg / g dry weight (Fig. 34).

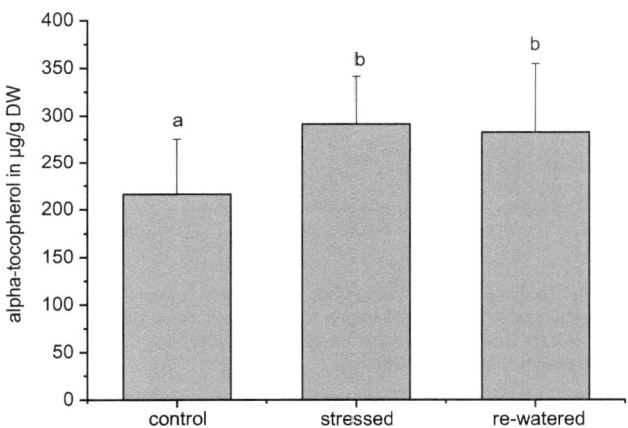

Figure 34: Total content of alpha-tocopherol in controls, stressed and re-watered plants. Significant differences between the samples were indicated by different letters. $P < 0.05$ analyzed by Kruskal-Wallis-ANOVA, n = 10, error bars show standard deviation, DW = dry weight.

4.3.3 Ascorbate

As Fig. 35 shows, there was no significant difference according to the ascorbate contents of controls compared to stressed plants but in both contents were significantly lower than in re-watered plants. Contents of total ascorbate were between 900 and 1,700 µg / g dry weight.

The proportional distribution of oxidized and reduced ascorbate in borage leaves of controls, stressed and re-watered plants were calculated relating to total ascorbate amounts and shown in Fig. 36. Reduced ascorbate was about 30 % in controls and stressed plants but (not significantly) increasing during re-watering to about 45 %.

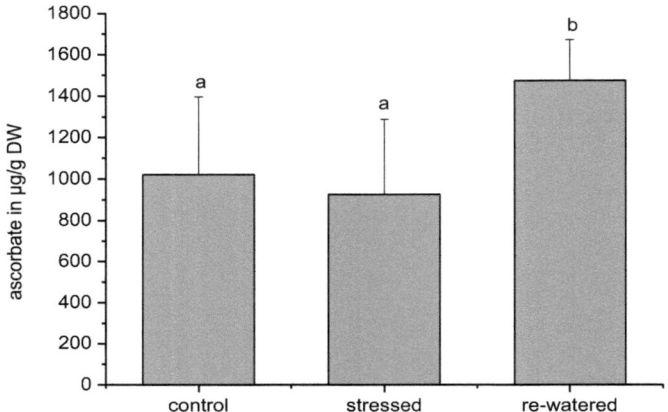

Figure 35: Total content of ascorbate in controls, stressed and re-watered plants. Significant differences between the samples were indicated by different letters. P < 0.05 analyzed by Kruskal-Wallis-ANOVA, n = 10, error bars show standard deviation, DW = dry weight.

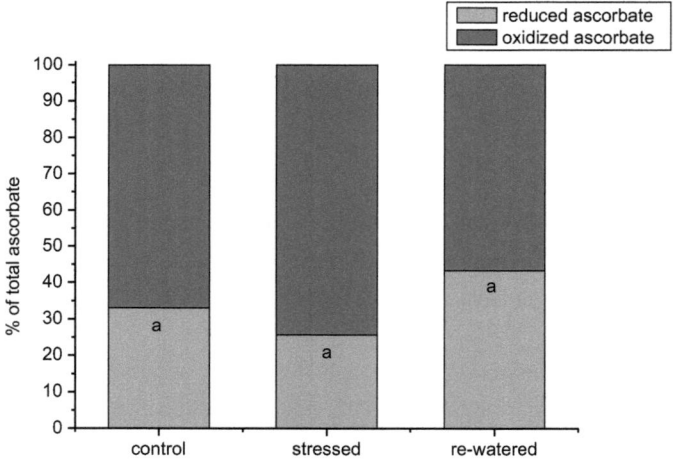

Figure 36: Percentage of oxidized and reduced ascorbate in controls, stressed and re-watered plants. Percentage was calculated relating to total ascorbate amounts. Significant differences between the samples are indicated by different letters. P < 0.05 analyzed by Kruskal-Wallis-ANOVA, n = 10.

4.3.4 Glutathione

Contents of total glutathione were between 2,500 and 5,500 nmol / g dry weight and significantly lower in both stressed and re-watered plants compared to controls. No significant differences were found between stressed and re-watered plants (*Fig. 37*).

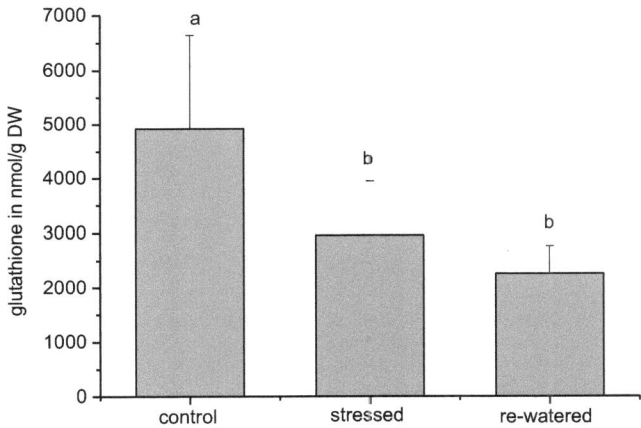

Figure 37: Total content of glutathione in controls, stressed and re-watered plants. Significant differences between the samples were indicated by different letters. P < 0.05 analyzed by Kruskal-Wallis-ANOVA, n = 10, error bars show standard deviation, DW = dry weight.

Figure 38 shows the proportional distribution of oxidized and reduced glutathione in controls, stressed and re-watered plants calculated relating to total glutathione amounts. Oxidized glutathione was between 60 - 70 % in controls and stressed plants and significantly lower (50 %) in re-watered plants. During recovery the percentage of oxidized glutathione was decreasing.

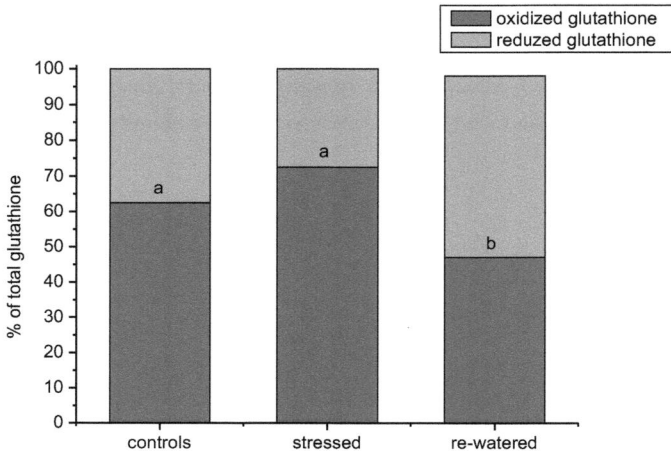

Figure 38: Percentage of oxidized and reduced glutathione in controls, stressed and re-watered plants. Percentage was calculated relating to total glutathione amounts. Significant differences between the samples are indicated by different letters. $P < 0.05$ analyzed by Kruskal-Wallis-ANOVA, n = 10.

4.3.5 Summary of the results of borage

Table 7 shows the significant changes of all examined substances as an overview and summary.

Table 7: Summary of significant changes in control, stressed and re-watered borage plants. ↑ = increase, ↓ = decrease, ↔ = no significant change, black arrows: compared to control plants, red arrows in brackets: compared to stressed plants.

Borage	stressed	re-watered	
chlorophyll a	↔	↔	(↔)
chlorophyll b	↔	↔	(↔)
α-carotene	↔	↑	(↑)
β-carotene	↔	↔	(↔)
neoxanthin	↔	↔	(↔)
violaxanthin	↔	↔	(↔)
antheraxanthin	↔	↔	(↔)
lutein+zeaxanthin	↔	↔	(↔)
chlorophyll a/b	↔	↔	(↔)
α-tocopherol	↑	↑	(↔)
ascorbate	↔	↑	(↑)
glutathione	↓	↓	(↔)

4.4 Discussion of the drought stress experiments

Plants can avoid drought by either maximizing water uptake or minimizing water loss (KOZLOWSKI & PALLARDY 2002). Apart from morphological changes, plants have developed a variety of physiological and biochemical processes (WANG & al. 2003, YIN & al. 2004) to deal with drought (LEI & al. 2006). Drought tolerant xerophyte species, for instance, possess great carotene and xanthophyll contents to protect their photosynthetic apparatus and maintain photochemical efficiency (VERES & al. 2006). The responses to a range of various stresses can underlie the phenomenon of priming (= previous stress exposure makes a plant more resistant to future exposure). Primed plants activate their defense responses faster and/or stronger to following pathogen attacks or abiotic stress. There is evidence that plants are adapt to modify their physiology and metabolism in response to prior experience. However, the mechanisms by which plants store information ("memory of plants") are hypothetical and unidentified until now (BRUCE & al. 2007). As there is a scarcity of studies dealing with recovery from water stress intensities (FLEXAS & al. 2006) our studies did not just focus on the status during stress but in particular on the status after. Therefore mild drought was used to prime our plants and the recovery phase could possibly give information about the plant status "memorizing" the stress.

Although depending on various factors like nature of the biological material (PINTEA & al. 2003), population (LEI & al. 2006), plant species and stress intensity (GALMES & al. 2007b, SIRCELJ & al. 2007) both, our results and a variety of studies, mostly showed a general response to stress by increasing tocopherol, ascorbate and pigment contents in photosynthetic tissues during drought (BARTOLI & al. 1999, GARCÍA-PLAZAOLA & BECERRIL 2000, BLOKHINA & al. 2003, LEI & al. 2006, SIRCELJ & al. 2007) and a decrease in the rate of photosynthetic CO_2 assimilation and transpiration (CAI & al. 2005, SOFO & al. 2005).

Changes in pigment concentrations seem to be strongly species-dependent because summer savory and borage showed no significant changes, in contrast nasturtium showed an increase. Results of examined apple trees (SIRCELJ & al. 2007) as well as chives (EGERT & TEVINI 2002) confirm our results for summer savory and borage. Furthermore, results of GALMES & al. (2007a) showed that chlorophyll contents did not significantly change during mild drought in most examined species (except in *D. ibicensis* contents of chlorophyll a and b were increased during mild drought stress) and therefore they suggest that adjusting chlorophyll may not be a major photo-protective response to water stress but can be a strong species-dependent feature. Chlorophyll contents of nasturtium changed similar to *D. ibicensis* (increase) (GALMES & al. 2007a) and therefore nasturtium is one of the species possessing this feature, while summer savory and borage are not (constant chlorophyll a and b content). On the other hand various studies showed that chlorophyll contents can also significantly decrease under drought (e.g. SEEL & al. 1992, MORAN & al. 1994, ASHRAF 2003, PARIDA & al. 2004, PAGTER & al. 2005, LEI & al. 2006, LI & al. 2008). The variations of total chlorophyll contents is as multifaceted as the one of the ratio of chlorophyll a to b (chl a/b). In borage and summer savory the ratio remained constant while in nasturtium it actually increased (stronger increase of chlorophyll a compared to chlorophyll b leads to higher chl a/b ratio). Similar results (as for borage and summer savory) were described in LEI & al. (2006) who also found a constant remaining ratio in *Populus* sp.. Changes in xanthophylls were significantly in nasturtium, but not as expected and known under high light conditions (e.g. decrease of violaxanthin, increase of zeaxanthin). In borage and summer savory for instance, no component of the xanthophyll cycle showed significant changes (exception: decrease of violaxanthin in summer savory). These results could be due to the fact that we were not able to examine zeaxanthin and lutein separately. Furthermore, measuring of the activity of enzymes involved in the xanthophyll-cycle could give more information to discuss our results.

According to various studies with different plant species and our results of nasturtium and borage plants α-tocopherol concentrations showed an increase in leaves during water deficiency (BARTOLI & al. 1999, GARCÍA-PLAZAOLA & BECERRIL 2000, MUNNÉ-BOSCH & ALEGRE 2000a and b, BLOKHINA & al. 2003). Indeed, attention should be paid to stress intensity as well as plant species, because our mild drought increased tocopherol contents in nasturtium and borage, but on the other hand no change was found in summer savory. SIRCELJ & al. (2007) showed, that mild drought did not affect tocopherol contents in apple trees but moderate drought increased them. Therefore it seems that mild drought for one species can be moderate or even severe for another one.

While pigments and glutathione contents of nasturtium, borage and summer savory plants decreased to or even below control contents after stopping of the drought and re-watering the plants (exception: pigments of borage showed no significant change) α-tocopherol contents doubled. As it is described in MUNNÉ-BOSCH (2005) changes in tocopherol concentrations are characterized by to phases – first there is an increase in tocopherol synthesis followed by loss of net tocopherol when the stress is too severe. These results suggest that increase of α-tocopherol seems to be a slower but long lasting effect in all three examined species while e.g. increase of pigments in nasturtium (especially chlorophyll a, β-carotene, neoxanthin, antheraxanthin and lutein+zeaxanthin) is a fast short time reaction of stress response to defend their photosynthetic apparatus from ROS.

No significant changes during mild drought were found in ascorbate contents in nasturtium as well as borage plants. These results fit with those of apple trees examined by SIRCELJ & al. (2007). However, summer savory showed an increase. Similar changes were found in *Populus* sp. (LEI & al. 2006) and *Cistus albidus* L. (JUBANHY-MARÍ & al. 2010). Interestingly, during re-watering ascorbate content increased significantly in summer savory and borage while remaining constant in nasturtium. Increase of ascorbate in summer savory and borage therefore seems to be a slower but long lasting effect (also cp. tocopherol).

Although glutathione contents significantly increased in nasturtium, in summer savory no change, in borage even a decrease was found. In apple trees (SIRCELJ & al. 2007). glutathione contents remained constant during drought stress too. However, nasturtium, summer savory and borage plants showed significantly decreases of glutathione contents after re-watering. Interestingly, the composition of glutathione (percentage oxidized and reduced) in re-watered plants changed towards lower oxidized levels in summer savory and borage, towards higher oxidized levels in nasturtium. All these results show the complexity of the ascorbate-glutathione-cycle. For further conclusions it would be necessary to measure enzyme activities and other components (like glutamate, cystein, glycine ...).

All these results show the importance of our examined antioxidants as well as enzymes and various other plant specific substances (not examined in our studies) e.g. essential oils in summer savory (increase under water stress (BAHER & al. 2002), phenolic compounds in borage (WETTASINGHE & al. 2001) and glucotropaeolin in nasturtium (increase during UV-B stress, SCHREINER & al. 2009) and underline in series once more how species dependent and multifaceted the plant defense network is.

As expected, drought had an impact on photosynthetic parameters in nasturtium. For instance it is well known that drought induces stomatal closure as our results and others showed (e.g. KOZLOWSKI & PALLARDY 2002, LUAN 2002, SCHROEDER & al. 2001, SIRCELJ & al. 2007), NEILL & al. 2008). The drawback of this stomata closure is the limitation of CO_2 fixation, which leads to an increase of intercellular CO_2 concentration (C_i) (significantly changes found in our stressed nasturtium plants) and may furthermore benefit the production of active oxygen species in the chloroplasts because of the excess of excitation of energy (ASADA 1999). In series, moderate drought in apple trees was finally leading to decreased C_i contents (SIRCELJ & al. 2007) because of lasting photosynthetic consumption and stomatal closure, in severe drought stress C_i increased again but net photosynthesis decreased. Therefore SIRCELJ & al. (2007) demonstrated that not CO_2 but a down-regulation or damage of the photosynthetic apparatus was limiting. In apple trees and some *Prunus* hybrids net photosynthesis and transpiration rates decreased during drought (SOFO & al. 2005, SIRCELJ & al. 2007) and increased after recovery to control levels (SOFO & al. 2005). This trend was also observed in assimilation rate of various herbaceous species, evergreen and semi-deciduous shrubs during drought and recovery (GALMES & al. 2007b). Although pigment concentrations and assimilation rate in re-watered nasturtium plants settled down again to control level, the other photosynthetic parameters (stomatal conductance, Ci and transpiration rate) did not significantly increase that high. These results suggest that examined pre-stressed nasturtium plants deal their stomatal regulation with great care to obviate further water loss. Observing the light response curves both, our results and various studies showed that net photosynthesis is strongly decreased during drought stress (KIRSCHBAUM 1988, FLEXAS & al. 2006). Remarkably, pre-stressed nasturtium plants used the light faster and a multiple more effective for assimilation in the low intensities, but showed a decrease in higher densities while control plants did not use light that effective but showed a constant increase.

In conclusion, components of the antioxidative network (of all examined species) and photosynthetic parameters (nasturtium) are modified due to mild drought. It is well known that enhancement of antioxidant defense in plants can increase tolerance to different stress factors (reviewed in WANG & al. 2003). Relating to drought this is also known as drought acclimation or drought hardening in many crop species (SRIVALLI & al. 2003, SELOTE & KHANNA-CHOPRA 2006). Therefore, we agree with JIANG & HUANG (2000) as well as with OH & al. (2009), not only to use drought pre-conditioning for tolerance increasing (JIANG & HUANG 2000) but also targeted for having an impact on contents of antioxidative substances in economic plants like it is shown in OH & al. (2009) for lettuce. However, it has to be acted with caution when using drought pre-conditioning against biotic stressors (like phytopathogens) because increased levels of ROS scavenging enzymes can abrogate the plants' ability to build up ROS for programmed cell death (reviewed in APEL & HIRT 2004). It is also important to bear in mind that not only the few substances examined by us, but also a great variety of others as well as different enzymes are involved in the regulation of the plant defense network (reviewed in WANG & al. 2003). Altogether, we agree with SIRCELJ & al. (2007) that factors like plant species, stress severity and various others affect changes of antioxidants during water deprivation. However, the multifaceted response of the antioxidant network in various plants during drought is confirming the facts that 1) plants in some respect seem to recollect the stress situations and hence as a learning effect adapt their metabolism (BRUCE & al. 2007) and 2) stress responses are species and stress intensity dependent (GALMES & al. 2007b, (SIRCELJ & al. 2007, CRUZ DE CARVALHO & al. 2010) and can not be standardized.

5. Drying and storage experiments

5.1 Results of nasturtium

Contents of various substances were examined in fresh controls (= CF), dried controls (= CD), dried plant material 6 months stored in tie bags (= 6MT) and 6 months stored in paper bags (= 6MP).

5.1.1 Pigments

Total content of chlorophylls were about 10,000 µg in fresh controls (CF), 7,000 µg in dried controls (CD), 3,500 µg in tie bag stored material and 4,000 µg / g dry weight in paper bag stored material. Chlorophyll contents were significantly decreasing during drying (to ¾ of fresh controls) and storage (to less than ½ of fresh controls). In nasturtium the loss of chlorophylls was significantly higher in tie bag compared to paper bag stored plant material (*Fig. 39*).

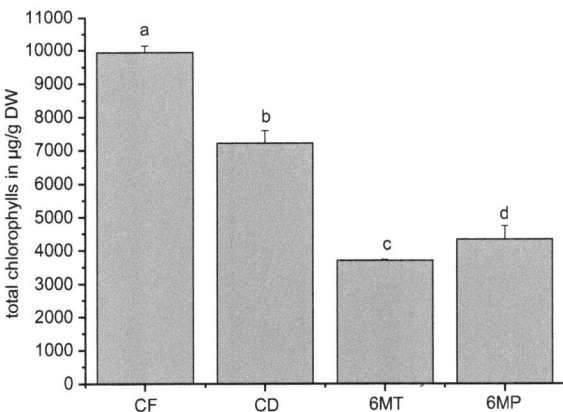

Figure 39: Total content of chlorophylls in fresh controls (= CF), dried controls (=CD), dried plant material 6 months stored in tie bags (= 6MT) and 6 months stored in paper bags (= 6MP). Significant differences between the samples were indicated by different letters. $P < 0.05$ analyzed by Kruskal-Wallis-ANOVA, n = 10, error bars show standard deviation, DW = dry weight.

Contents of total carotenoids were about 700 and 2,000 µg / g dry weight and were significantly different in all four variants. Fresh controls showed the highest carotenoid contents, a slight decrease was found in dried controls and 6 months stored material in tie bags showed the lowest contents (*Fig. 40*).

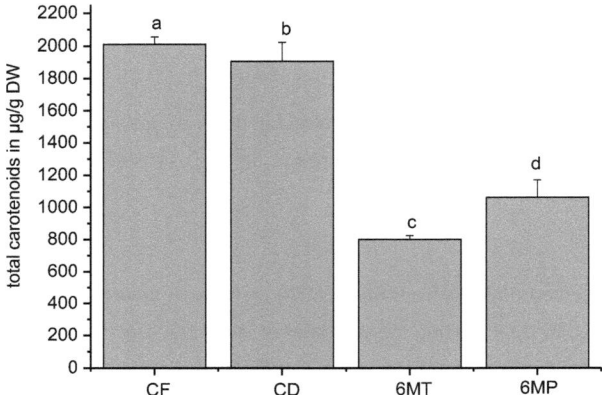

Figure 40: Total content of carotenoids in fresh controls (= CF), dried controls (=CD), dried plant material 6 months stored in tie bags (= 6MT) and 6 months stored in paper bags (= 6MP). Significant differences between the samples were indicated by different letters. P < 0.05 analyzed by Kruskal-Wallis-ANOVA, n = 10, error bars show standard deviation, DW = dry weight.

The ratio of chlorophyll a / b was between 2.0 and 3.5 and was significantly different in all four variants. The significantly highest ratio was found in the fresh controls, the lowest in 6 months paper bag stored material (*Fig. 41*).

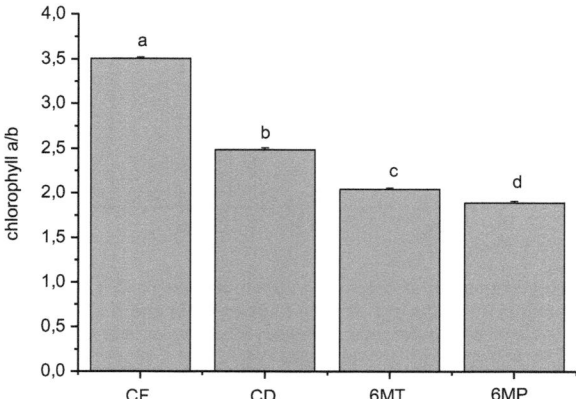

Figure 41: Ratio of chlorophyll a / b for fresh controls (= CF), dried controls (=CD), dried plant material 6 months stored in tie bags (= 6MT) and 6 months stored in paper bags (= 6MP). Significant differences between the samples were indicated by different letters. P < 0.05 analyzed by Kruskal-Wallis-ANOVA, n = 10, error bars show standard deviation, DW = dry weight.

5.1.2 Tocopherol

Tocopherol contents were significantly lowest in the fresh controls and increasing during drying and storage. Highest content of tocopherol was found in 6 months tie bag stored plant material (*Fig. 42*).

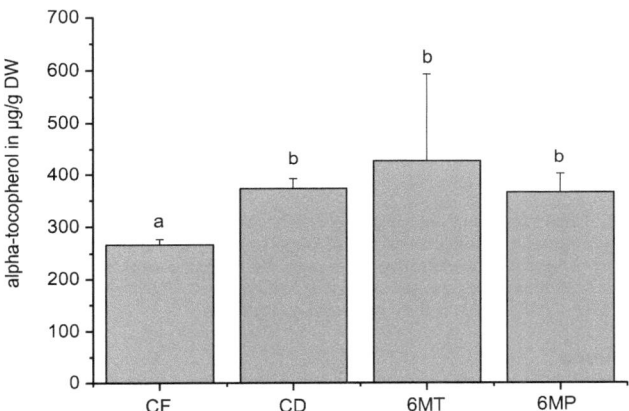

Figure 42: Total content of alpha-tocopherol in fresh controls (= CF), dried controls (=CD), dried plant material 6 months stored in tie bags (= 6MT) and 6 months stored in paper bags (= 6MP). Significant differences between the samples were indicated by different letters. P < 0.05 analyzed by Kruskal-Wallis-ANOVA, n = 10, error bars show standard deviation, DW = dry weight.

5.1.3 Ascorbate

As *Fig. 43* shows was the significantly highest ascorbate content in fresh controls and a decrease found in dried material. As the contents were so low no detection was possible in both, 6 months tie and paper bag stored material.

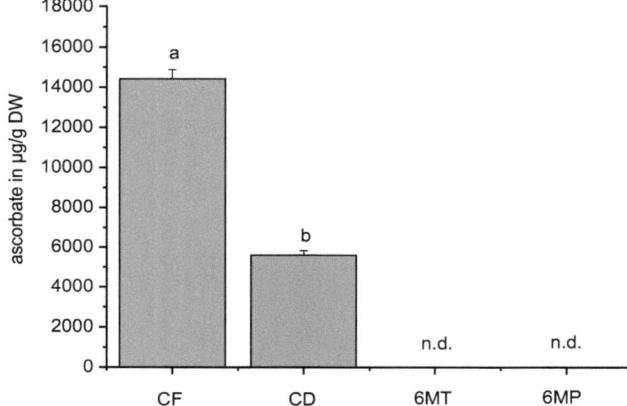

Figure 43: Total content of ascorbate in fresh controls (= CF), dried controls (=CD), dried plant material 6 months stored in tie bags (= 6MT) and 6 months stored in paper bags (= 6MP). Significant differences between the samples were indicated by different letters. P < 0.05 analyzed by Kruskal-Wallis-ANOVA, n = 10, error bars show standard deviation, DW = dry weight, n.d. = no detection possible.

5.1.4 Glutathione

Contents of total glutathione were between 3,500 and 7,500 nmol / g dry weight and significantly higher in dried controls and stored material compared to fresh controls. No significant differences were found between 6 months tie and paper bag stored material (*Fig. 44*).

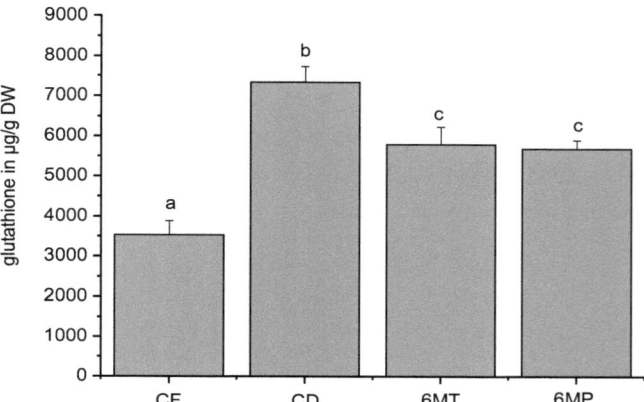

Figure 44: Total content of glutathione in fresh controls (= CF), dried controls (=CD), dried plant material 6 months stored in tie bags (= 6MT) and 6 months stored in paper bags (= 6MP). Significant differences between the samples were indicated by different letters. P < 0.05 analyzed by Kruskal-Wallis-ANOVA, n = 10, error bars show standard deviation, DW = dry weight.

5.1.5 Summary of the results of nasturtium

Table 8 shows the significant changes of all examined substances as an overview and summary.

Table 8: Summary of significant changes in dried, 6 months tie and paper bag stored nasturtium plants. Dried controls were compared to fresh controls, stored material to dried controls. ↑ = increase, ↓ = decrease, ↔ = no significant change, n.d. = no detection possible, black arrows: compared to control plants, red arrows in brackets: compared to stressed plants.

Nasturtium	dried	tie bags	paper bags
chlorophylls	↓	↓	↓ (↑)
carotenoids	↓	↓	↓ (↑)
chlorophyll a/b	↓	↓	↓ (↓)
α-tocopherol	↑	↔	↔ (↔)
ascorbate	↓	n.d.	n.d. n.d.
glutathione	↑	↓	↓ (↔)

5.2 Results of summer savory

Contents of various substances were examined in fresh controls (= CF), dried controls (= CD), dried plant material 3 months stored in tie bags (= 3MT) and 3 months stored in paper bags (= 3MP).

5.2.1 Pigments

Total content of chlorophylls were about 6,000 µg in fresh controls, 5,500 µg in dried controls, 2,000 µg in tie bag stored material and 3,500 µg / g dry weight in paper bag stored material. Chlorophyll contents were not significantly decreasing during drying but during storage. In summer savory the loss of chlorophylls was significantly higher in tie bag compared to paper bag stored plant material (*Fig. 45*).

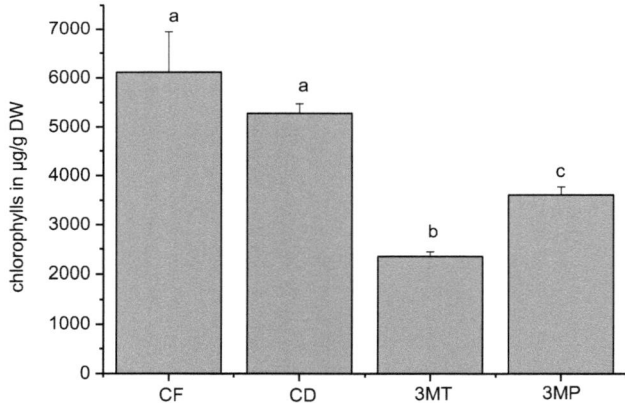

Figure 45: Total content of chlorophylls in fresh controls (= CF), dried controls (=CD), dried plant material 3 months stored in tie bags (= 3MT) and 3 months stored in paper bags (= 3MP). Significant differences between the samples were indicated by different letters. $P < 0.05$ analyzed by Kruskal-Wallis-ANOVA, n = 10, error bars show standard deviation, DW = dry weight.

Contents of total carotenoids were about 300 and 1,100 µg / g dry weight and were significantly different in all four variants. Fresh controls showed the highest carotenoid contents, a significant decrease was found in dried controls and 3 months stored material in tie bags showed the lowest contents (*Fig. 46*).

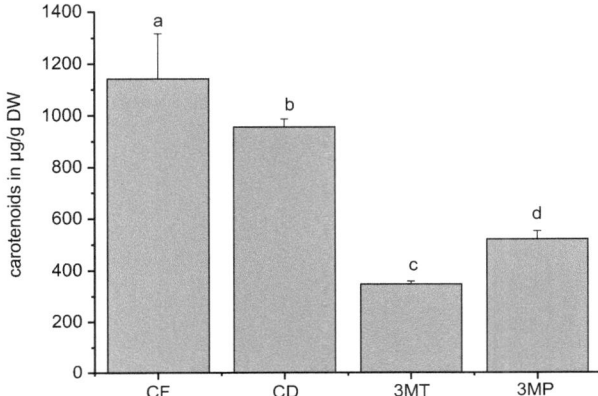

Figure 46: Total content of carotenoids in fresh controls (= CF), dried controls (=CD), dried plant material 3 months stored in tie bags (= 3MT) and 3 months stored in paper bags (= 3MP). Significant differences between the samples were indicated by different letters. $P < 0.05$ analyzed by Kruskal-Wallis-ANOVA, n = 10, error bars show standard deviation, DW = dry weight.

The ratio of chlorophyll a / b was between 2.3 and 3.2 and was significantly different in all four variants. The significantly highest ratio was found in the fresh controls, the lowest in 3 months tie bag stored material (*Fig. 47*).

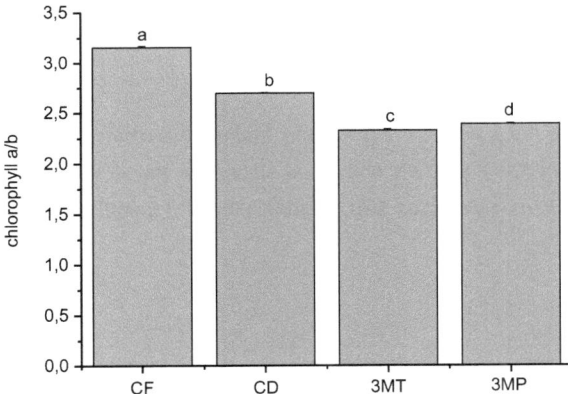

Figure 47: Ratio of chlorophyll a / b for fresh controls (= CF), dried controls (=CD), dried plant material 3 months stored in tie bags (= 3MT) and 3 months stored in paper bags (= 3MP). Significant differences between the samples were indicated by different letters. $P < 0.05$ analyzed by Kruskal-Wallis-ANOVA, n = 10, error bars show standard deviation, DW = dry weight.

5.2.2 Tocopherol

Tocopherol contents were significantly highest in the fresh controls and decreasing during drying and storage by more than 50 %. Significantly lowest content of tocopherol was found in 3 months tie bag stored plant material (*Fig. 48*).

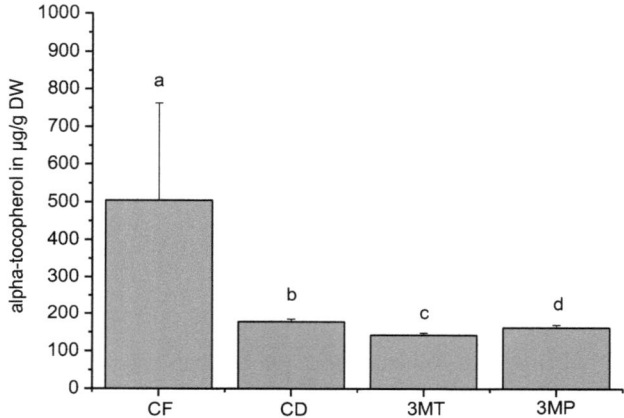

Figure 48: Total content of alpha-tocopherol in fresh controls (= CF), dried controls (=CD), dried plant material 3 months stored in tie bags (= 3MT) and 3 months stored in paper bags (= 3MP). Significant differences between the samples were indicated by different letters. P < 0.05 analyzed by Kruskal-Wallis-ANOVA, n = 10, error bars show standard deviation, DW = dry weight.

5.2.3 Ascorbate

As *Fig. 49* shows there was the significantly highest ascorbate content in fresh controls (about 2,000 µg/g DW) and an enormous decrease found in dried material. As the contents were so low no detection was possible in both, 3 months tie and paper bag stored material.

Drying and storage experiments

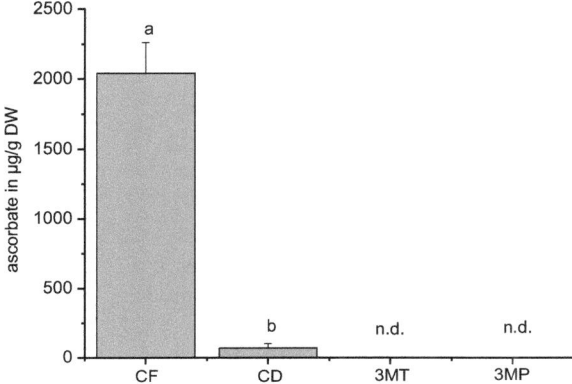

Figure 49: Total content of ascorbate in fresh controls (= CF), dried controls (=CD), dried plant material 3 months stored in tie bags (= 3MT) and 3 months stored in paper bags (= 3MP). Significant differences between the samples were indicated by different letters. P < 0.05 analyzed by Kruskal-Wallis-ANOVA, n = 10, error bars show standard deviation, DW = dry weight, n.d. = no detection possible.

5.2.4 Glutathione

Contents of total glutathione were between 500 and 2,000 nmol / g dry weight and significantly highest in fresh controls compared to dried controls and both storage variants. No significant differences were found between dried controls and 3 months paper bag stored material as well as between both storage variants (*Fig. 50*).

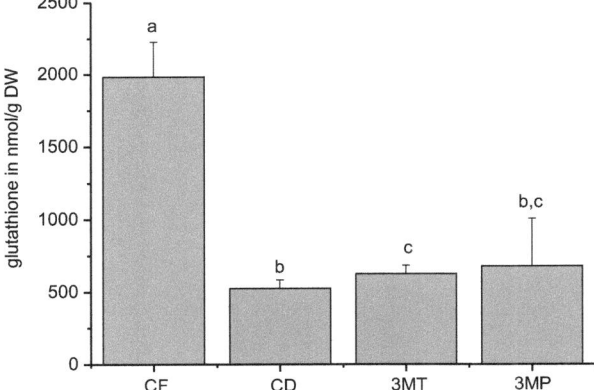

Figure 50: Total content of glutathione in fresh controls (= CF), dried controls (=CD), dried plant material 3 months stored in tie bags (= 3MT) and 3 months stored in paper bags (= 3MP). Significant differences between the samples were indicated by different letters. P < 0.05 analyzed by Kruskal-Wallis-ANOVA, n = 10, error bars show standard deviation, DW = dry weight.

5.2.5 Summary of the results of summer savory

Table 9 shows the significant changes of all examined substances as an overview and summary.

Table 9: Summary of significant changes in dried, 3 months tie and paper bag stored summer savory plants. Dried controls were compared to fresh controls, stored material to dried controls. ↑ = increase, ↓ = decrease, ↔ = no significant change, n.d. = no detection possible, black arrows: compared to control plants, red arrows in brackets: compared to stressed plants.

Summer savory	dried	tie bags	paper bags
chlorophylls	↔	↓	↓ (↑)
carotenoids	↓	↓	↓ (↑)
chlorophyll a/b	↓	↓	↓ (↑)
α-tocopherol	↓	↓	↓ (↑)
ascorbate	↓	n.d.	n.d. n.d.
glutathione	↓	↑	↔ (↔)

5.3 Results of borage

Contents of various substances were examined in fresh controls (= CF), dried controls (= CD), dried plant material 3 months stored in tie bags (= 3MT) and 3 months stored in paper bags (= 3MP).

5.3.1 Pigments

Total content of chlorophylls were about 9,000 µg in fresh controls, 8,000 µg in dried controls, 7,000 µg in tie bag stored material and 4,000 µg / g dry weight in paper bag stored material. Chlorophyll contents were significantly decreasing during drying and storage significantly lowest in paper bag stored material (*Fig. 51*).

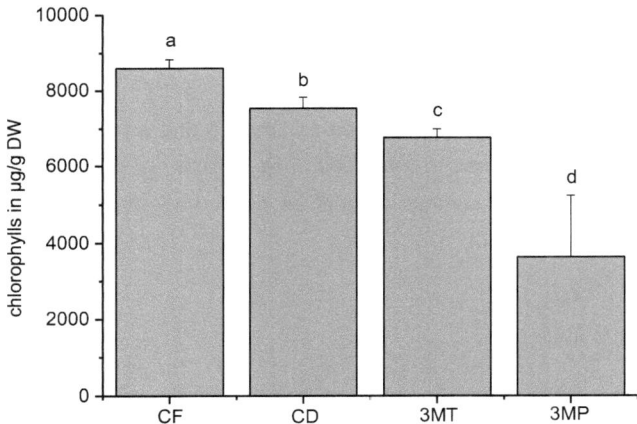

Figure 51: Total content of chlorophylls in fresh controls (= CF), dried controls (=CD), dried plant material 3 months stored in tie bags (= 3MT) and 3 months stored in paper bags (= 3MP). Significant differences between the samples were indicated by different letters. P < 0.05 analyzed by Kruskal-Wallis-ANCVA, n = 10, error bars show standard deviation, DW = dry weight.

Contents of total carotenoids were about 500 and 1,800 µg / g dry weight and were significantly different in all four variants. Fresh controls showed the highest carotenoid contents, a significant decrease was found in dried controls and 3 months stored material. Paper bag stored material showed the lowest contents (*Fig. 52*).

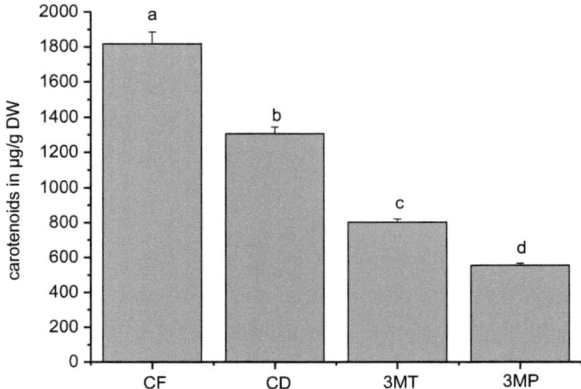

Figure 52: Total content of carotenoids in fresh controls (= CF), dried controls (=CD), dried plant material 3 months stored in tie bags (= 3MT) and 3 months stored in paper bags (= 3MP). Significant differences between the samples were indicated by different lowercase letters. P < 0.05 analyzed by Kruskal-Wallis-ANOVA, n = 10, error bars show standard deviation, DW = dry weight.

The ratio of chlorophyll a / b was between 3.0 and 3.5 and was significantly lower in tie bag stored material compared to fresh and dried controls as well as paper bag stored material. Fresh and dried controls as well as paper bag stored material showed no significant difference (*Fig. 53*).

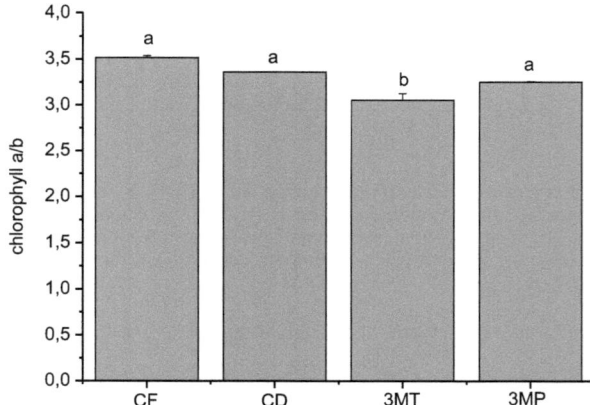

Figure 53: Ratio of chlorophyll a / b for fresh controls (= CF), dried controls (=CD), dried plant material 3 months stored in tie bags (= 3MT) and 3 months stored in paper bags (= 3MP). Significant differences between the samples were indicated by different lowercase letters. P < 0.05 analyzed by Kruskal-Wallis-ANOVA, n = 10, error bars show standard deviation, DW = dry weight.

5.3.2 Tocopherol

Tocopherol contents were significantly highest in the fresh controls and decreasing during drying and storage by more than 60 %. No significant diccerence was found between dried and 3 months tie bag as well as paper bag stored plant material (*Fig. 54*).

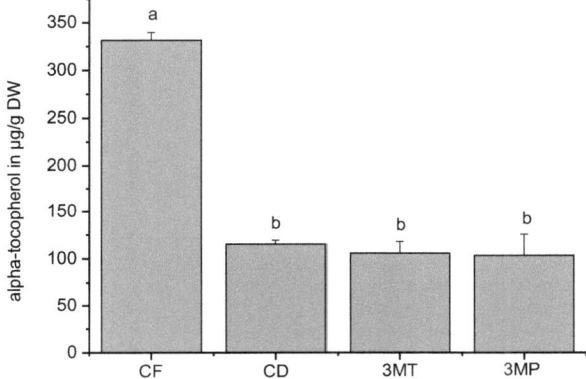

Figure 54: Total content of alpha-tocopherol in fresh controls (= CF), dried controls (=CD), dried plant material 3 months stored in tie bags (= 3MT) and 3 months stored in paper bags (= 3MP). Significant differences between the samples were indicated by different letters. $P < 0.05$ analyzed by Kruskal-Wallis-ANOVA, n = 10, error bars show standard deviation, DW = dry weight.

5.3.3 Ascorbate

As *Fig. 55* shows there was the only detectable ascorbate content in fresh controls about 5,500 µg/g DW. In all other variants no detection was possible.

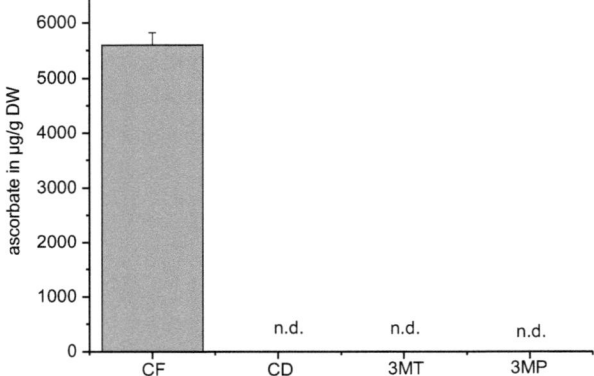

Figure 55: Total content of ascorbate in fresh controls (= CF), dried controls (=CD), dried plant material 3 months stored in tie bags (= 3MT) and 3 months stored in paper bags (= 6MP). Error bars show standard deviation, DW = dry weight, n.d. = no detection possible.

5.3.4 Glutathione

Contents of total glutathione were between 100 and 2,300 nmol / g dry weight and significantly highest in fresh controls. Between dried controls and paper bag stored material significant differences were found too. In 3 months tie bag stored material detection of glutathione was not possible (*Fig. 56*).

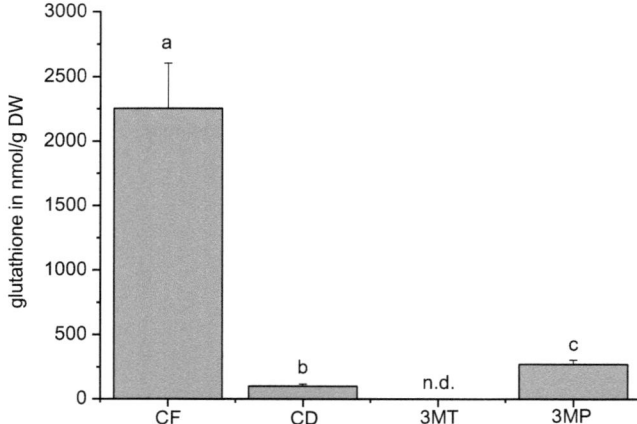

Figure 56: Total content of glutathione in fresh controls (= CF), dried controls (=CD), dried plant material 3 months stored in tie bags (= 3MT) and 3 months stored in paper bags (= 3MP).Significant differences between the samples were indicated by different letters. P < 0.05 analyzed by Kruskal-Wallis-ANOVA, n = 10, error bars show standard deviation, DW = dry weight, n.d. = no detection possible.

5.3.5 Summary of the results of borage

Table 10 shows the significant changes of all examined substances as an overview and summary.

Table 10: Summary of significant changes in dried, 3 months tie and paper bag stored borage plants. Dried controls were compared to fresh controls, stored material to dried controls. ↑ = increase, ↓ = decrease, ↔ = no significant change, n.d. = no detection possible, black arrows: compared to control plants, red arrows in brackets: compared to stressed plants.

Borage	dried	tie bags	paper bags	
chlorophylls	↓	↓	↓	(↓)
carotenoids	↓	↓	↔	(↓)
chlorophyll a/b	↔	↓	↔	(↑)
α-tocopherol	↓	↔	↔	(↔)
ascorbate	n.d.	n.d.	n.d.	n.d.
glutathione	↓	n.d.	↑	n.d.

5.4 Discussion of drying and storage experiments

Lipid peroxidation is an important reaction leading to deterioration of food during storage and processing (YAGI 1987). Especially the quality of lipid-containing products, the prolongation of their storage time and their optimum stabilization is directly associated with the addition of suitable antioxidants (reviewed in YANISHLIEVA & al. 2006). Therefore, to increase the quality of food the use of antioxidants is important to prevent or delay oxidative deterioration during processing and storage (COSIO & al. 2006). Furthermore, research is not only focussed on food processing and post-harvest storage but also on cultivar development and production practices, which could maintain or even improve antioxidant contents and activity in plant originated material (KALT & al. 1999). In general, it is well known that many factors such as temperature and pH of the media, processing treatment (e.g. heat) and storage can have a strong influence on the activity and concentration of antioxidants (GAZZANI & al. 1998). Synthetic antioxidants have been used as food additives for a long time, but safety concerns and reports on their involvement in chronic diseases have restricted their use in food. Therefore, international attention has been directed towards natural antioxidants mainly from plant sources (CEBALLOS & FERNÁNDEZ 2000), and the development and utilization of more effective antioxidants of natural origin (GÜLÇIN & al. 2002, OKTAY & al. 2003). Furthermore, the reduction of synthetic chemicals in food preservation (and cosmetics) is desired. Especially common culinary herbs, spices and aromatic plants could be attractive natural alternatives and are one of the most important targets to search for natural antioxidants from the safety´ point of view (CONNER & BEUCHAT 1984, DEANS & RITCHIE 1987, CONNER 1993, BEUCHAT 1994, YANISHLIEVA & al. 2006).

Plant originated antioxidants are not only supplements for food (enhancing food quality) but also used in the pure form as spices or seasoning herbs and it must be kept in mind that both forms can be affected by various processes. For example, a lot of herbs and spices are not only used fresh but also dried. Therefore it is important to know that composition of phytochemicals can be significantly changed due to enzymatic processes during drying of fresh plant material (JAMBOR & CZOSNOWSKA 2002) and due to degradation during storage.

In our studies we examined nasturtium, summer savory and borage plant material after drying and different conditions of storage to detect decompositions of various antioxidants. For drying we used a drying hurdle ensuring constant temperature of 35°C, good ventilation and protection from light because traditional sun drying for example (which is the cheapest and most accessible means of food preservation) can cause considerable carotenoid destruction due to the photo-response of these pigments. Interestingly, studies of VISHALAKSHI DEVI (2003) and REDDY & al. (2005) showed that controlled oven drying with 50°C resulted in maximum nutrient retention (e.g. for ascorbic acid, α-tocopherol, β-carotene and glutathione). For our plant samples however, 35°C were high enough and in the range of temperature recommended by DACHLER & PELZMANN (1999) for gentle drying of plants containing essential oils. Also storage of the plant material was with constant room temperature, in bags permeable to air and in the dark. Protection from light during storage to diminish carotenoid decomposition is e.g. recommended by RAJU & al. (2007).

Carotenoids are long known as labile substances (DAVIES 1976) and are often the victim of oxidative degradation which can be accelerated by e.g. light, heat or enzymes (RODRIGUEZ-AMAYA 2002, RODRIGUEZ & RODRIGUEZ-AMAYA 2007). This is due to the conjugated double system of carotenoids (isomerization), which is not only important for their functions but is also responsible for their susceptibility to oxidation (DE OLIVIERA & RODRIGUEZ-AMAYA 2007). Furthermore, carotenoids can be easily destroyed during drying processes (DAVIES 1976). Degradation of chlorophylls during thermal processes is the main cause of green color loss (which is a problem for food industry). This is due to the Mg^{2+} loss and therefore the origin of pheophytins leads to unprofitable brown colors (ESKIN 1990). Pigment stability varies between different food even if the same storage/processing conditions are used and therefore optimum conditions during preparation, storage etc. are different, too (RODRIGUEZ-AMAYA 2009) and must be identified for each herb. Spices and herbs should accordingly be tested in the actual food under realistic conditions prior practical use in food industry (YANISHLIEVA & al. 2006).

After drying in nasturtium, summer savory and borage plants contents of pigments (chlorophylls and carotenoids) were significantly decreased by 5 – 30 % (exception: chlorophylls in summer savory remained constant). Our results were confirmed by CAPECKA & al. (2005) who also showed significant decreases for peppermint, lemon balm and oregano. In lemon myrtle (*Backhousia citriodora*) colors of leaves were changed unfavorable and especially high temperature drying caused pigment degradation (BUCHAILLOT & al. 2009); in Tencha (Japanese green tea) total content of chlorophylls decreased during drying at high temperature (80–200°C), too (KOHATA & al. 2001). DAOOD & al. (2006) showed that carotenoids in ground paprika react sensible in different ways not only depending on temperature and type of carotenoid but also on sort of paprika spice (pungent or non-pungent). During storage most plants showed a degradation of chlorophyll content (ROURA & al. 2000, MOREIRA & al. 2003, PRABHU & BARRETT 2009). According to temperature sensitivity, NEGI & ROY (2004) found out that total chlorophyll contents of fresh amaranth and fenugreek leaves decreased with increasing temperature and decreasing humidity of the storage room. On the other hand, NWUFO (1994) reported that storage temperature had no significant effect on chlorophyll degradation in *Telfairia occidentalis* and *Pterocarpus soyauxii*. In our studies a decrease of chlorophylls and carotenoids in dried material of all examined species and both storage types was found during storage at room temperature (about – 50 % of the fresh content! - exception: carotenoids in borage remained constant in paper bags). These results are confirmed by LU & al. (2010), who showed that the level of total chlorophyll content in *Salicornia bigelovii* decreased during storage at different temperatures, which is similar to broccoli's storage (COSTA & al. 2005). In ground paprika total carotenoid contents decreased up to about 50 % during 3 month storage at 15-20°C in the dark (MARKUS & al. 1999), in amaranth and fenugreek also great losses of β-carotenoids were found during storage (NEGI & ROY 2004). On the other hand, GALLARDO-GUERRERO & al. (2005) showed, that during 1 year's storage at 15°C in the dark the total pigment content remained constant in virgin olive oil, but the composition changed (increased pheophytinization, decreased chlorophyll a). This could be due to the fact that pigments vary in their chemical composition and are therefore differential in their sensitivity to degradation. Also the consistency of (plant) material (e.g. leaves, oil, powder...) seems to be crucial. Interestingly, in our studies there is decomposition of pigments in both, paper and special tie bags (equipped with foil - as it is used for storage of teas). Actually, the simpler and cheaper paper bags had higher pigments contents after storage compared to tie bags. This fact suggests that tie bags are possibly useful for the protection of essential oils but are not required for protection of pigments.

Ascorbate (vitamin C) is well examined because of both, its importance for human health and as one of the major antioxidants in plant species. It is water soluble, quantitatively the predominant antioxidant in plant cells and is found in all subcellular compartments, cytoplasma as well as the apoplast (see e.g. JIMÉNEZ & al. 1997 and 1998, VANACKER & al. 1998, SMIRNOFF 2000, KOLLIST & al. 2001, VAN HOVE & al. 2001, NOCTOR & al. 2002, TAKAHAMA 2004). Various studies showed the sensitivity of ascorbate due to water loss caused by drying. For example storage (and in series water loss) of spinach leaves at room temperature for only three days caused a loss of ascorbate of 90% (reviewed in DIPLOCK & al. 1998). In general, various studies (e.g. NEGI & ROY 2004, DAOOD & al. 2006) showed that the degradation of ascorbic acid is depending on temperature and (storage) duration and follows first order kinetics (GUTZEIT & al. 2008). Therefore, fruits and vegetables showed a gradual decrease in ascorbic acid content as storage temperature and/or duration increased (various in LEE & KADER 2000, GIL-IZQUIERDO & al. 2001, PIGA & al. 2003, AN & al. 2006, DEROSSI & al. 2010). The loss of ascorbic acid content was most probably dominated by the presence of catalysts and oxidase enzymes (such as polyphenol oxidase) catalysing the oxidation especially at high temperature (GIL-IZQUIERDO & al. 2001, MAO & al. 2006, LI & al. 2008). In our studies ascorbate was significantly decreased in all three plant species (–60 to – 100 %!) and even veered towards zero in borage (no detection possible) during drying and storage. These results are confirmed by CAPECKA & al. (2005) who showed a reduction of ascorbate in lemonbalm and peppermint of about -90 % caused by drying. In all examined species it was no longer possible even to detect ascorbate in stored material. PRABHU & BARRETT (2009) showed that in African leafy vegetables (*Cassia tora, Corchorus tridens*) even after frozen storage (-18°C) for 1 or 2 months no more detection of ascorbic acid was possible. Fitting to dependence on temperature and/or duration for example GUTZEIT & al. (2008) found out that total ascorbic acid in sea buckthorn juice degraded stored at room temperature for 7 days by about 20%, stored at elevated temperature (40°C) by about 45%. LU & al. (2010) showed that the slowest decreasing rate for the ascorbic acid contents of *Salicornia bigelovii* (sea asparagus) appeared at 2°C and the sharpest decreasing rate was at 25°C. However, in studies of KALT & al. (1999) fruits (berries) showed (if at all) minimal loss of ascorbate, processed pineapple products (e.g. juice, sorbet and jam) a significant decrease (UCKIAH & al. 2009) during storage. These results let us assume that ascorbate contents and stability depend on various factors (e.g. consistence of (plant) material; cp. pigments) and that dried and stored herbs (compared to some fruits) drop out as potential source for ascorbate due to its very fast degradation.

The (water-soluble) glutathione showed increases, decreases, no changes or even "no detection possible" during drying and storage in our plant species. Unfortunately, we found no authors dealing with glutathione contents during drying and storage for comparison. However, our results let us assume, that complex (degradation-) processes occur in plant material during drying and storage showing once more the dependency of various (unknown) factors.

Tocopherols have aromatic rings with methyl substituents and fully saturated isoprenoid tails (DELLAPENNA & LAST 2006), are therefore highly lipophilic and operative in membranes (e.g. cell membrane, intra- and extracellulary (RODRIGO & al. 2007) and lipoproteins (DIPLOCK & al. 1998). In comparison with other lipophilic antioxidants, α-tocopherol is probably the most efficient in the lipid phase (DIPLOCK & al. 1998) and meets human vitamin E requirements best (reviewed in TRABER & ATKINSON 2007) because of its highest bioavailability (RODRIGO & al. 2007) and the highest vitamin E activity (DELLAPENNA & LAST 2006). The most important antioxidant function appears to be the inhibition of lipid peroxidation (DIPLOCK & al. 1998), and therefore α-tocopherol is used as food additive (DIPLOCK 1992, LANGSETH 1995). In general, (similar to other antioxidants) the degradation of tocopherols seems to be depended of temperature and (storage) duration and follows first order kinetics (DAOOD & al. 2006, LAVELLI & al. 2006, HIDALGO & al. 2009, SABLIOV & al. 2009). In our studies, α-tocopherol content was affected during drying and storage, it was decreased in summer savory and borage (– 70 %!), but interestingly increased significantly in nasturtium plants (about + 15 %!) during drying. In rapeseed drying with different temperatures caused decreases in tocopherol contents of -4 to -11 % (GAWRYSIAK-WITULSKA & al. 2009). During storage in paper and tie bags α-tocopherol content remained constant in nasturtium and borage plants, in summer savory it decreased (again paper bags showed significantly higher contents compared to tie bags). Studies of GAWRYSIAK-WITULSKA & al. (2009) showed that during storage the level of tocopherol in rapeseeds was also significantly reduced which is similar to stored ground paprika (MARKUS & al.1999) and soybean oil (PLAYER & al. 2006). These results showed that contents of tocopherol were affected by drying and storage but far from the extent shown for e.g. ascorbate. In the exceptional case of nasturtium drying could even be used for enhancing α-tocopherol content. Furthermore, contents in nasturtium (as well as borage) plant materials are not further degraded during storage, so that these herbs (with remaining contents between 100 and 400 µg g^{-1} DW) can supplement the human tocopherol household.

In conclusion, our studies showed that drying and storage caused great losses of antioxidants in summer savory, borage and nasturtium (with a few exceptions). In general, degradation of various antioxidative substances is depending on temperature and duration (during both, drying and storage) and is mostly following first order kinetics (various authors, see above). Although, the higher the temperature, the greater the antioxidant degradation, even frozen storage can lead to significant losses of ascorbate and pigment contents (LISIEWSKA & KMIECIK 1997, PRABHU & BARRETT 2009). In studies of CHOHAN & al. (2008) the total antioxidant capacity of culinary herbs and thus the impact on the dietary antioxidant contribution of these herbs were affected by storage, too. Surprisingly, in our studies selection of special storage packages did not lead to better results, tie bags were as unsuitable as simple paper bags for protecting antioxidants during storage. However, not only the substances examined by us, but also various others are responsible for the antioxidative properties of herbs and food. In culinary herbs for example, various studies suggest, that the polyphenolic compounds may be the major contributors to antioxidants (CROWELL 1999, HALVORSEN & al. 2002, DRAGLAND & al. 2003, HALVORSEN & al. 2006). Therefore, CAPECKA & al. (2005) supposed that regardless of a significant decrease in ascorbic acid and carotenoid contents after drying of plant material, (fresh and dried) lemon balm, oregano and peppermint are rich sources of antioxidants, in particular from the group of phenolic compounds. As our studies showed it should be kept in mind that nasturtium (because of the increase after drying) could be a good source for tocopherol and glutathione. Furthermore, MILOS & al. (2000) and ZHENG & WANG (2001) suggested that not only the level of antioxidants but also a synergy occurring between them and the other plant constituents might influence the differences in the antioxidant ability of plant extracts (CAPECKA & al. 2005) and even if e.g. a product lacks antioxidant properties, it may still possess other biological properties important for the defense mechanisms of the cells against chemical or biological aggression (CAILLET & al. 2007).

6. Processing experiments

6.1 Results of nasturtium

Contents of various substances were examined in (see also M&M):
- fresh controls (=CF)
- cut material (=C)
- cut and 10 min rested material (=CR)
- cut and 5 min boiled material (=C5)
- cut and 20 min boiled material (=C20)
- Cut, rested and 5 min boiled material (=CR5)
- Cut, rested and 20 min boiled (=CR20)
- Dried control (=CD)
- Dried and 5 min boiled (=DB5)
- Dried and 20 min boiled (=DB20)

6.1.1 Pigments

Total content of chlorophylls was between 7,000 µg and 16,000 µg / g dry weight. After drying and processing an increase of total chlorophyll contents were found. Contents of chlorophylls showed a significant increase after boiling for both, 5 and 20 minutes compared to fresh controls. However, 20 min boiled fresh material showed lower chlorophyll contents than 5 min boiled material (*Fig. 57*). *Figure 58* shows the total content of chlorophylls in fresh controls (=CF), cut material (=C), cut and 5 min boiled material (=CB5) and cut and 20 min boiled material (=CB20). Chlorophyll content showed a significant increase in cut, cut and 5 min boiled and cut and 20 minutes boiled material compared to fresh controls. A significant decrease was found between cut and 5 minutes boiled and cut and 20 minutes boiled material, no significant difference between cut and cut and 20 minutes boiled material. Total content of chlorophylls in fresh controls (=CF), cut material (=C), cut and let rested material (=CR), cut, let rested material boiled for 5 min (=CR5) and for 20 min (=CR20) are shown in *Fig. 59*.

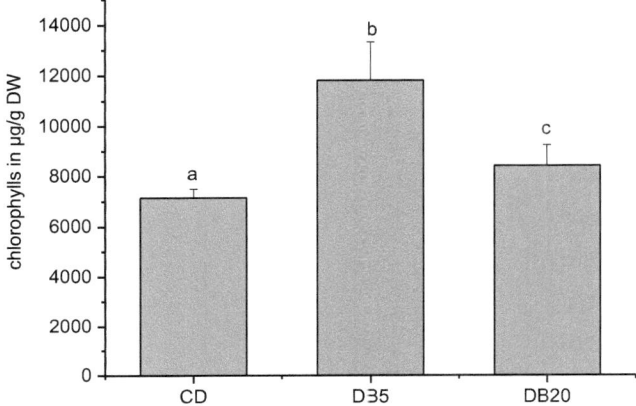

Figure 57: Total content of chlorophylls in dried controls (=CD), dried and 5 min boiled (=DB5) and dried and 20 min boiled (=DB20). Significant differences between the samples were indicated by different letters. P < 0.05 analyzed by Kruskal-Wallis-ANOVA, n = 10, error bars show standard deviation, DW = dry weight.

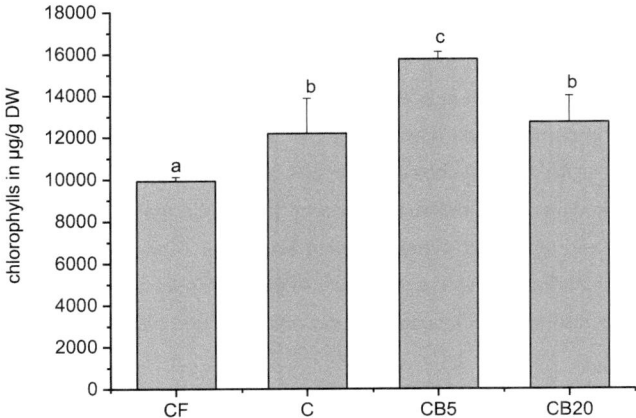

Figure 58: Total content of chlorophylls in fresh controls (=CF), cut material (=C), cut and 5 min boiled material (=CB5) and cut and 20 min boiled material (=CB20). Significant differences between the samples were indicated by different letters. P < 0.05 analyzed by Kruskal-Wallis-ANOVA, n = 10, error bars show standard deviation, DW = dry weight.

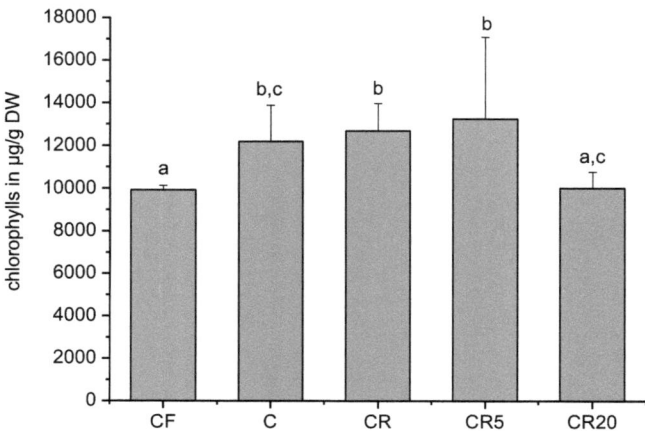

Figure 59: Total content of chlorophylls in fresh controls (=CF), cut material (=C), cut and let rested material (=CR), cut, let rested material boiled for 5 min (=CR5) and for 20 min (=CR20). Significant differences between the samples were indicated by different letters. P < 0.05 analyzed by Kruskal-Wallis-ANOVA, n = 10, error bars show standard deviation, DW = dry weight.

Contents of total carotenoids of dried plant material (between 1,800 and 2,500 µg/g DW) showed significant increase after boiling for 5 and 20 minutes, highest content was found in 5 min boiled material (*Fig. 60*). Total carotenoid contents were significantly increased after cutting, and boiling for 5 and 20 minutes compared to fresh controls. Contents were between 2,000 and more than 4,000 µg/g DW and significantly highest in 5 min boiled material (*Fig. 61*). As *Fig. 62* shows, there was no significant change of total carotenoid contents in cut and let rested material compared to only cut material but again a significant increase in cut and let rested material during boiling for 5 minutes (highest content). 20 minutes boiled material showed significantly higher contents compared to fresh controls but no change compared to cut material or cut and let rested material.

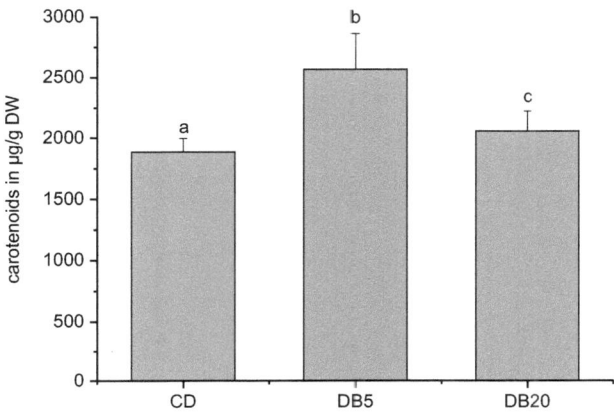

Figure 60: Total content of carotenoids in dried controls (=CD), dried and 5 min boiled (=DB5) and dried and 20 min boiled (=DB20) Significant differences between the samples were indicated by different lowercase letters. $P < 0.05$ analyzed by Kruskal-Wallis-ANOVA, $n = 10$, error bars show standard deviation, DW = dry weight.

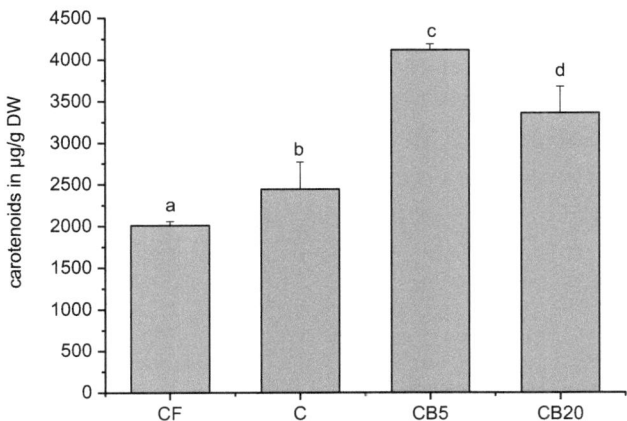

Figure 61: Total content of carotenoids in fresh controls (=CF), cut material (=C), cut and 5 min boiled material (=CB5) and cut and 20 min boiled material (=CB20). Significant differences between the samples were indicated by different letters. $P < 0.05$ analyzed by Kruskal-Wallis-ANOVA, $n = 10$, error bars show standard deviation, DW = dry weight.

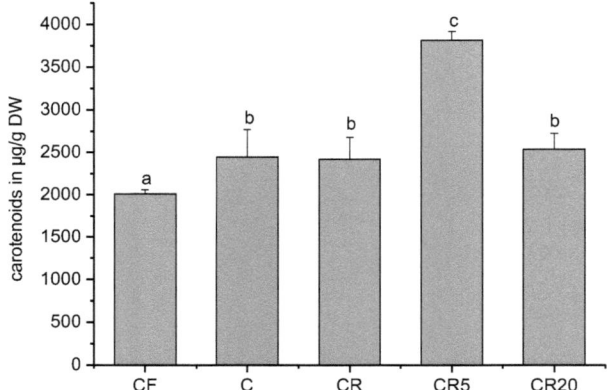

Figure 62: Total content of carotenoids in fresh controls (=CF), cut material (=C), cut and let rested material (=CR), cut, let rested material boiled for 5 min (=CR5) and for 20 min (=CR20). Significant differences between the samples were indicated by different letters. P < 0.05 analyzed by Kruskal-Wallis-ANOVA, n = 10, error bars show standard deviation, DW = dry weight.

As *Fig. 63* shows, the ratio of chlorophyll a / b (between 2.5 and 3.0) was not significantly changing in dried and both, 5 and 20 minutes boiled material compared to dried controls. Also no significant changes between fresh controls, cut material and cut and both, 5 and 20 minutes boiled material were found (*Fig. 64*). Fresh material showed a higher ratio of chlorophyll a / b (about 3.5) compared to dried material (about 2.5 – 3.0) (*Fig. 63 – 65*). Ratio of chlorophyll a/b in fresh controls, cut material, cut and let rested material, cut, let rested material boiled for 5 min and for 20 min were not changing significantly (*Fig. 65*).

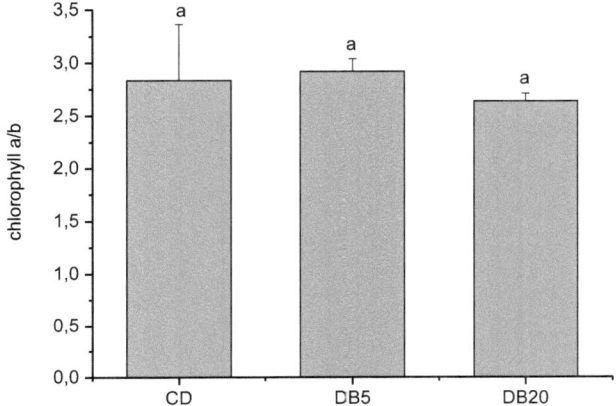

Figure 63: Ratio of chlorophyll a/b in dried controls (=CD), dried and 5 min boiled (=DB5) and dried and 20 min boiled (=DB20). Same letters mean no significant differences. P < 0.05 analyzed by Kruskal-Wallis-ANOVA, n = 10, error bars show standard deviation, DW = dry weight.

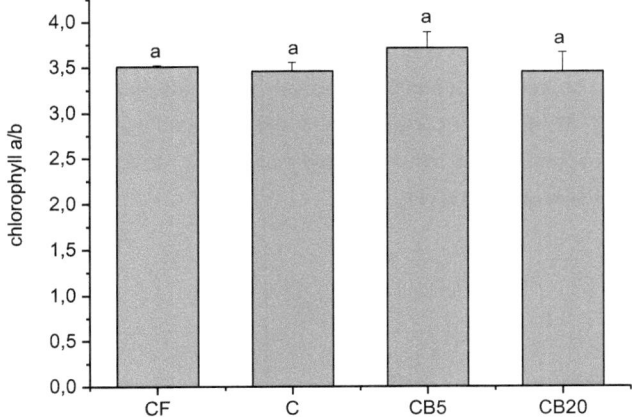

Figure 64: Ratio of chlorophyll a/b in fresh controls (=CF), cut material (=C), cut and 5 min boiled material (=CB5) and cut and 20 min boiled material (=CB20). Same letters mean no significant differences. P < 0.05 analyzed by Kruskal-Wallis-ANOVA, n = 10, error bars show standard deviation, DW = dry weight.

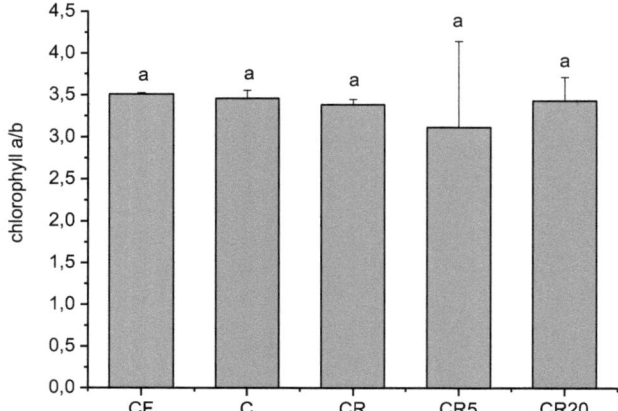

Figure 65: Ratio of chlorophyll a/b in fresh controls (=CF), cut material (=C), cut and let rested material (=CR), cut, let rested material boiled for 5 min (=CR5) and for 20 min (=CR20). Same letters mean no significant differences. $P < 0.05$ analyzed by Kruskal-Wallis-ANOVA, n = 10, error bars show standard deviation, DW = dry weight.

Calculated in % of fresh controls dried material showed a decrease of – 25%, dried and 5 minutes boiled material an increase of +20% and 20 minutes boiled material a decrease of -10% of total pigment contents (*Fig. 66*). An increase of pigment contents was found in cut material (+20%) as well as cut and 5 respectively 20 minutes boiled material (about +70% and +35%) (*Fig. 67*). Pigment contents in cut and let rested material showed an increase by +25%, cut, let rested and 5 minutes boiled material by +45% and cut, let rested and 20 minutes boiled material by +5% (*Fig. 68*).

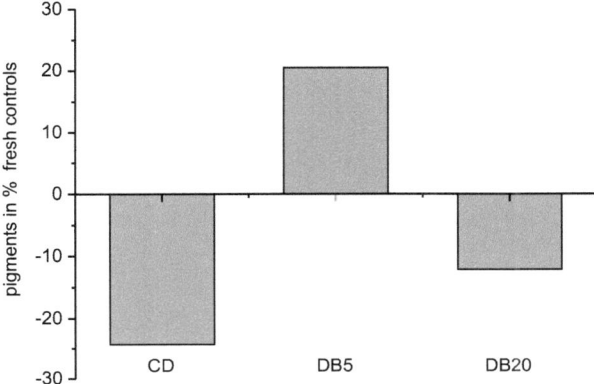

Figure 66: Increase/decrease of pigment contents in dried controls (=CD), dried and 5 min boiled (=DB5) and dried and 20 min boiled (=DB20) calculated in % of fresh controls. n = 10.

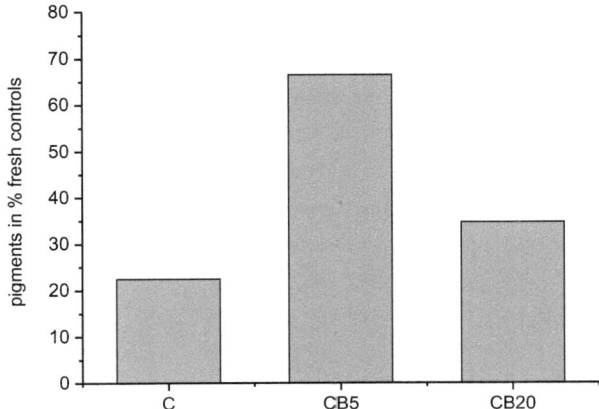

Figure 67: Increase of pigment contents in cut material (=C), cut and 5 min boiled material (=CB5) and cut and 20 min boiled material (=CB20). n = 10.

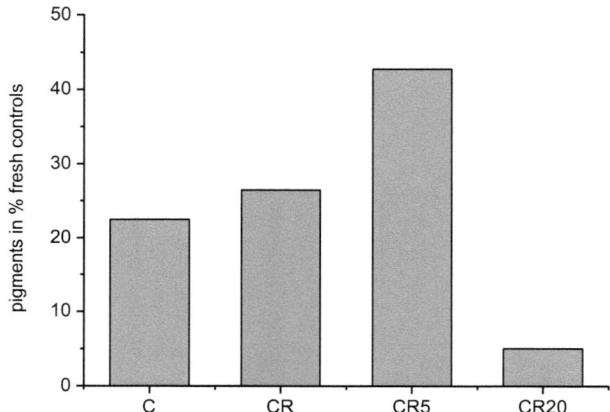

Figure 68: Increase of pigment contents in cut material (=C), cut and let rested material (=CR), cut, let rested material boiled for 5 min (=CR5) and for 20 min (=CR20). n = 10.

6.1.2 Tocopherol

Contents of alpha-tocopherol (about 350 µg/g DW) showed a significant decrease after boiling for 5 minutes compared to dried controls. No changes were found after boiling for 20 minutes (*Fig. 69*). As *Fig. 70* shows there were no significant differences in tocopherol contents between fresh controls and cut material (about 250 µg/g DW), but a significant increase after both, boiling for 5 and 20 minutes (450 – 500 µg/g DW). No difference was found between the two boiling variants. Tocopherol contents of cut and cut and afterwards let rested material (250 µg/g DW) showed no significant differences compared to fresh controls. Again an significant increase was found in 5 and 20 minutes boiled material (about 400 µg/g DW) (*Fig. 71*).

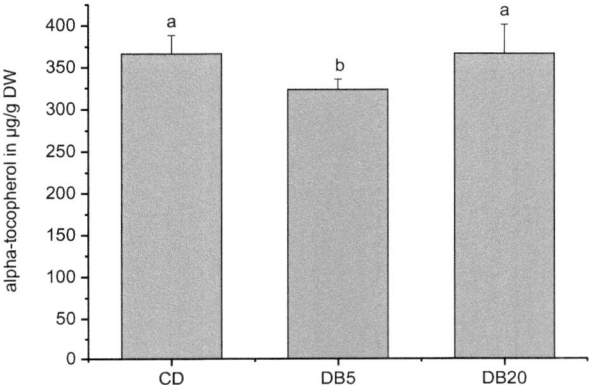

Figure 69: Total content of alpha-tocopherol in dried controls (=CD), dried and 5 min boiled (=DB5) and dried and 20 min boiled (=DB20). Significant differences between the samples were indicated by different letters. $P < 0.05$ analyzed by Kruskal-Wallis-ANOVA, n = 10, error bars show standard deviation, DW = dry weight.

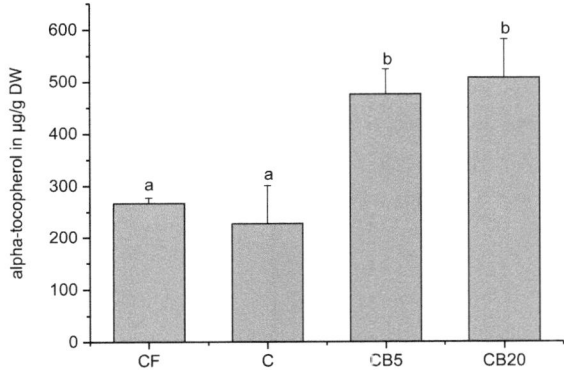

Figure 70: Total content of alpha-tocopherol in fresh controls (=CF), cut material (=C), cut and 5 min boiled material (=CB5) and cut and 20 min boiled material (=CB20). Significant differences between the samples were indicated by different letters. $P < 0.05$ analyzed by Kruskal-Wallis-ANOVA, n = 10, error bars show standard deviation, DW = dry weight.

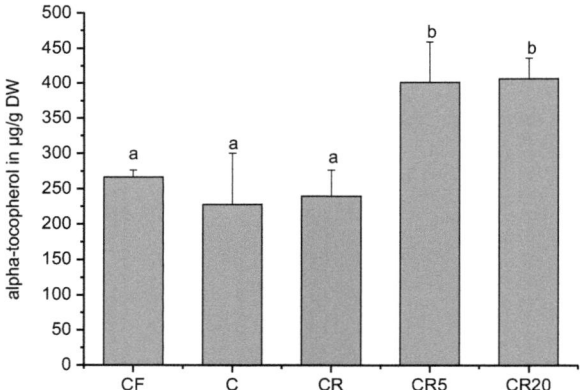

Figure 71: Total content of tocopherol in fresh controls (=CF), cut material (=C), cut and let rested material (=CR), cut, let rested material boiled for 5 min (=CR5) and for 20 min (=CR20). Significant differences between the samples were indicated by different letters. P < 0.05 analyzed by Kruskal-Wallis-ANOVA, n = 10, error bars show standard deviation, DW = dry weight.

Calculated in % of fresh controls dried material showed an increase of about +40%, dried and 5 minutes boiled material of +20% and 20 minutes boiled material +40% of total pigment contents (*Fig. 72*). An decrease of pigment contents was found in cut material about -20%) and increase in cut and 5 respectively 20 minutes boiled material (about +80% and +90%) (*Fig. 73*). Tocopherol contents in cut and let rested material decreased by -10%, cut, let rested and 5 minutes as well as 20 minutes boiled material increased by +50% (*Fig. 74*).

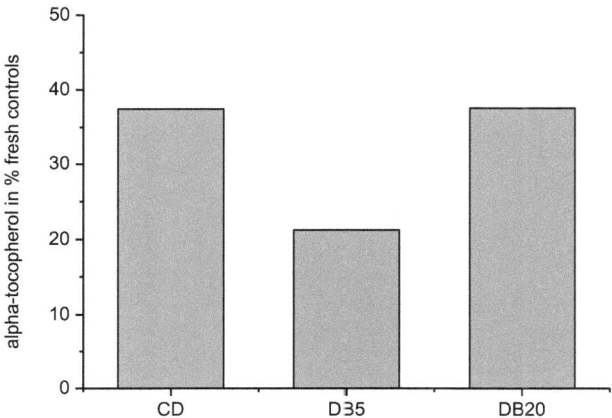

Figure 72: Increase of tocopherol contents in dried controls (=CD), dried and 5 min boiled (=DB5) and dried and 20 min boiled (=DB20) calculated in % of fresh controls. n = 10.

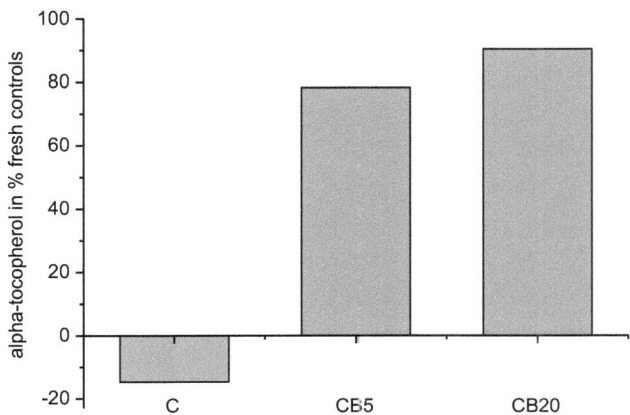

Figure 73: Increase/decrease of alpha-tocopherol in cut material (=C), cut and 5 min boiled material (=CB5) and cut and 20 min boiled material (=CB20) calculated in % of fresh controls. n = 10.

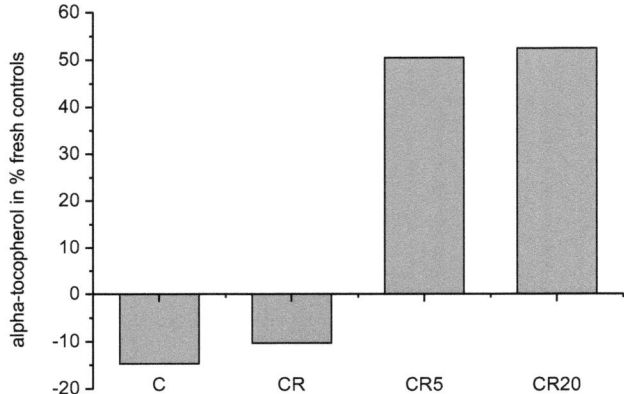

Figure 74: Increase/decrease of tocopherol in cut material (=C), cut and let rested material (=CR), cut, let rested material boiled for 5 min (=CR5) and for 20 min (=CR20) calculated in % of fresh controls. n = 10.

6.1.3 Ascorbate

Contents of ascorbate (about 5,500 µg/g DW) showed a significant decrease after boiling for both, 5 and 20 minutes compared to dried controls (100 µg/g DW). Boiling for 20 minutes showed a slightly significant increase of ascorbate compared to 5 minutes boiled material (*Fig. 75*). As *figure 76* shows there were no significant differences in ascorbate contents between fresh controls and cut material (14,000 – 16,000 µg/g DW), but a significant decrease after both, boiling for 5 and 20 minutes (1,500 and 2,000 µg/g DW). Again 20 minutes boiled material showed significantly higher contents compared to 5 min boiled material (*Fig. 76*). Ascorbate contents of cut and cut and afterwards let rested material (14,000 -16,000 µg/g DW) showed no significant differences compared to fresh controls. Again a significant decrease was found in 5 and 20 minutes boiled material (about 2,000 µg/g DW), 20 minutes boiled material showed by trend but not significantly higher contents compared to 5 minutes boiled material (*Fig. 77*).

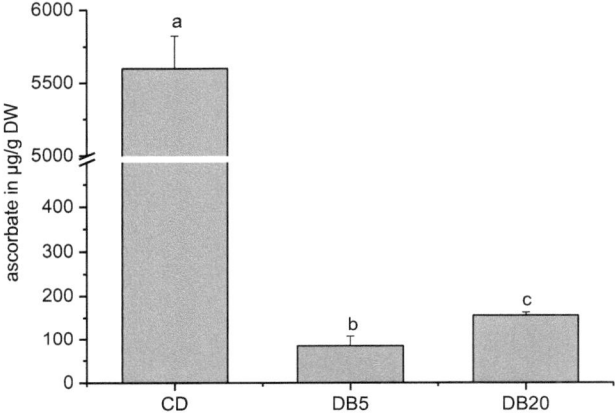

Figure 75: Total content of ascorbate in dried controls (=CD), dried and 5 min boiled (=DB5) and dried and 20 min boiled (=DB20). Significant differences between the samples were indicated by different letters. $P < 0.05$ analyzed by Kruskal-Wallis-ANOVA, n = 10, error bars show standard deviation, DW = dry weight.

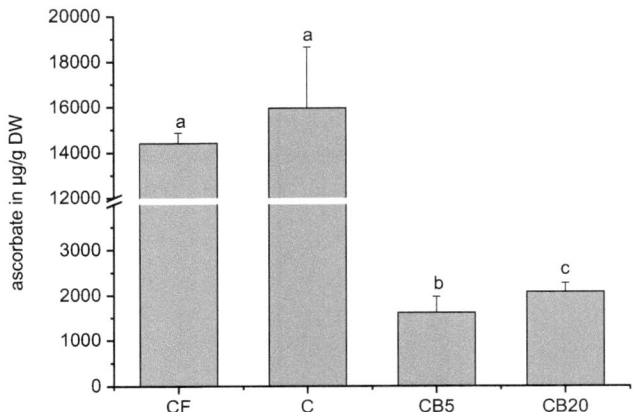

Figure 76: Total content of ascorbate in fresh controls (=CF), cut material (=C), cut and 5 min boiled material (=CB5) and cut and 20 min boiled material (=CB20). Significant differences between the samples were indicated by different letters. $P < 0.05$ analyzed by Kruskal-Wallis-ANOVA, n = 10, error bars show standard deviation, DW = dry weight.

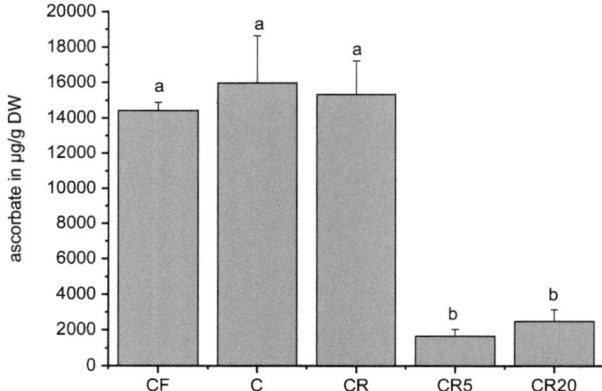

Figure 77: Total content of ascorbate in fresh controls (=CF), cut material (=C), cut and let rested material (=CR), cut, let rested material boiled for 5 min (=CR5) and for 20 min (=CR20). Significant differences between the samples were indicated by different letters. P < 0.05 analyzed by Kruskal-Wallis-ANOVA, n = 10, error bars show standard deviation, DW = dry weight.

Calculated in % of fresh controls dried material showed an decrease of about -60%, dried and 5 as well as 20 minutes boiled material nearly -100% of total ascorbate contents (*Fig. 78*). An increase of ascorbate contents was found in cut material about +10%) and decrease in cut and 5 respectively 20 minutes boiled material (about -90% and -80%) (*Fig. 79*). Ascorbate contents in cut and cut as well as let rested material increased by +5 and +10%, cut, let rested and 5 minutes as well as 20 minutes boiled material decreased by about -90% (*Fig. 80*).

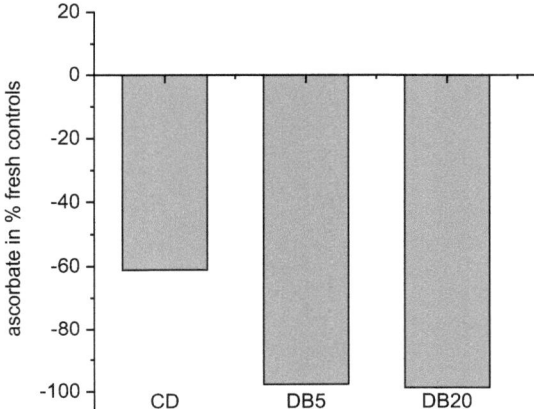

Figure 78: Decrease of ascorbate contents in dried controls (=CD), dried and 5 min boiled (=DB5) and dried and 20 min boiled (=DB20) calculated in % of fresh controls. n = 10.

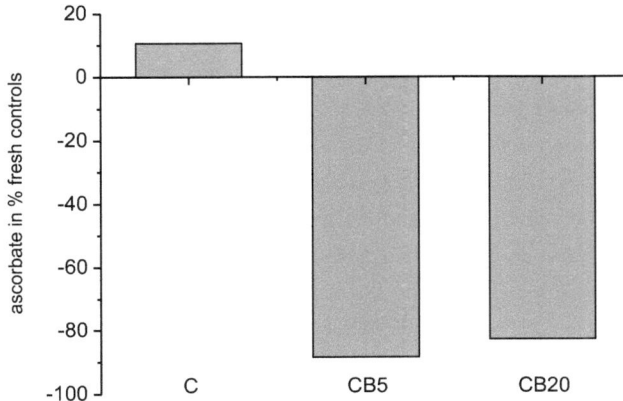

Figure 79: Increase/decrease of ascorbate in cut material (=C), cut and 5 min boiled material (=CB5) and cut and 20 min boiled material (=CB20) calculated in % of fresh controls. n = 10.

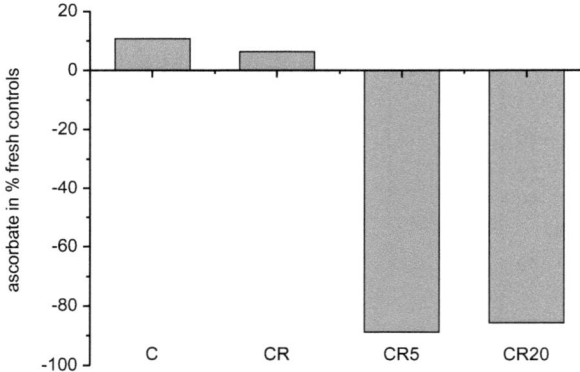

Figure 80: Increase/decrease of ascorbate in cut material (=C), cut and let rested material (=CR), cut, let rested material boiled for 5 min (=CR5) and for 20 min (=CR20) calculated in % of fresh controls. n = 10.

6.1.4 Glutathione

Contents of glutathione (about 7,000 nmol/g DW) were significantly decrease after boiling for both, 5 and 20 minutes compared to dried controls (800 – 1,000 nmol/g DW). Boiling for 20 minutes showed no significant change of glutathione content compared to 5 minutes boiled material (*Fig. 81*). As *figure 82* shows there were no significant differences in glutathione contents between fresh controls and cut material (about 3,500 nmol/g DW), but a significant decrease after both, boiling for 5 and 20 minutes (400 – 500 nmol/g DW). 20 minutes boiled material showed no significant difference compared to 5 min boiled material (*Fig. 82*). Glutathione contents of cut and cut and afterwards let rested material (about 3,500 nmol/g DW) showed no significant differences compared to fresh controls. Again a significant decrease was found in 5 and 20 minutes boiled material (about 300 – 400 nmol/g DW), 20 minutes boiled material showed by trend but not significantly higher contents compared to 5 minutes boiled material (*Fig. 83*).

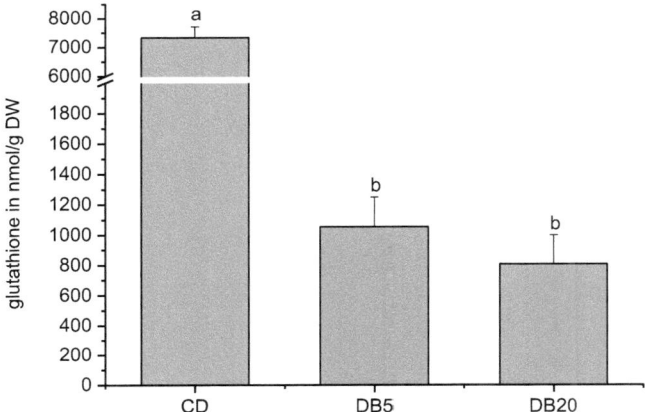

Figure 81: Total content of glutathione in dried controls (=CD), dried and 5 min boiled (=DB5) and dried and 20 min boiled (=DB20). Significant differences between the samples were indicated by different letters. P < 0.05 analyzed by Kruskal-Wallis-ANOVA, n = 10, error bars show standard deviation, DW = dry weight.

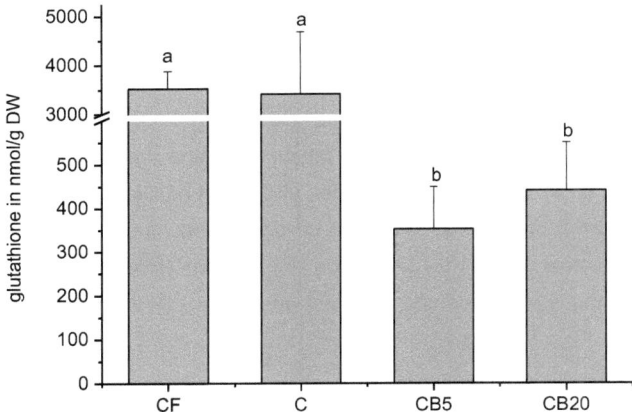

Figure 82: Total content of glutathione in fresh controls (=CF), cut material (=C), cut and 5 min boiled material (=CB5) and cut and 20 min boiled material (=CB20). Significant differences between the samples were indicated by different letters. P < 0.05 analyzed by Kruskal-Wallis-ANOVA, n = 10, error bars show standard deviation, DW = dry weight.

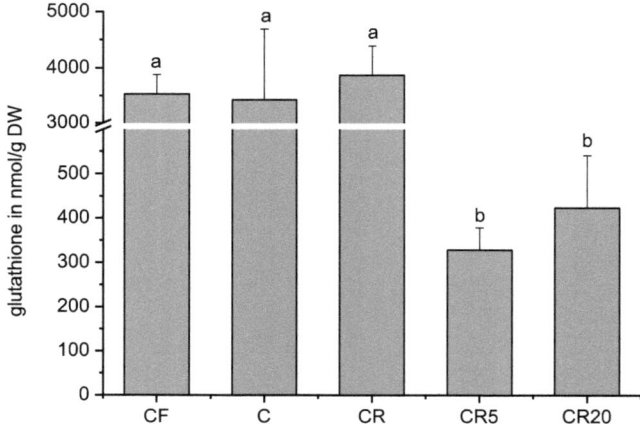

Figure 83: Total content of glutathione in fresh controls (=CF), cut material (=C), cut and let rested material (=CR), cut, let rested material boiled for 5 min (=CR5) and for 20 min (=CR20). Significant differences between the samples were indicated by different letters. P < 0.05 analyzed by Kruskal-Wallis-ANOVA, n = 10, error bars show standard deviation, DW = dry weight.

Figure 84 shows the proportional distribution of oxidized and reduced glutathione in dried controls, dried and 5 min boiled and dried and 20 min boiled material calculated relating to total glutathione amounts. Oxidized glutathione was about 90% in dried controls and (not significantly) lower after boiling for 5 minutes (about 80%) and significantly after boiling for 20 minutes (about 70%). Percentage of oxidized and reduced glutathione in fresh controls, cut material, cut and 5 min as well as 20 min boiled material, cut and let rested material, cut, let rested material boiled for 5 min and for 20 min (calculated relating to total glutathione amounts) was between 50 and 65% and not significantly changing (*Fig. 85 – 86*).

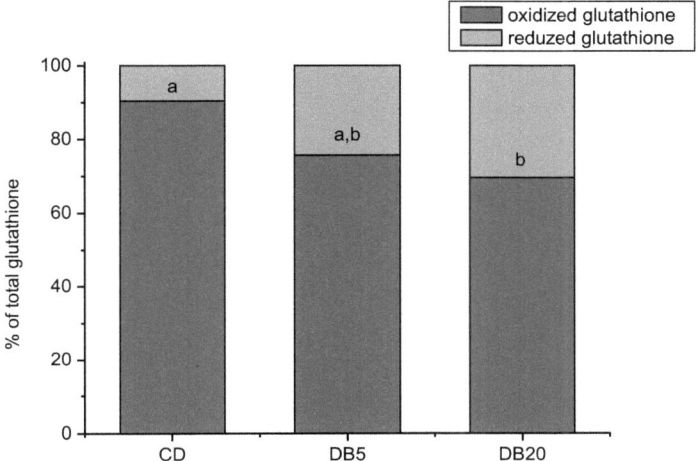

Figure 84: Percentage of oxidized and reduced glutathione in dried controls (=CD), dried and 5 min boiled (=DB5) and dried and 20 min boiled (=DB20). Percentage was calculated relating to total glutathione amounts. Significant differences between the samples were indicated by different letters. P < 0.05 analyzed by Kruskal-Wallis-ANOVA, n = 10.

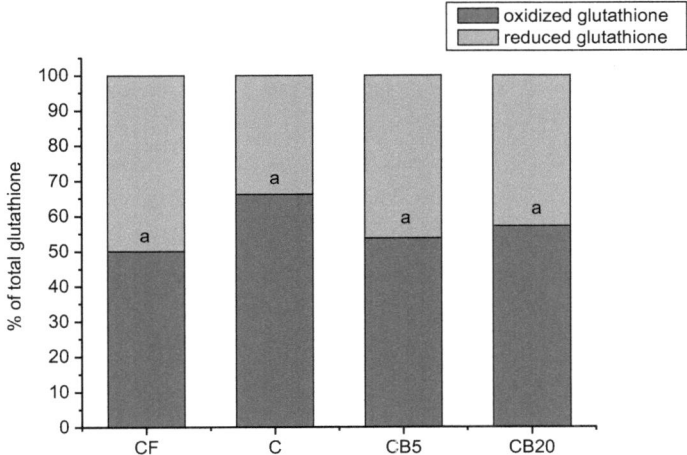

Figure 85: Percentage of oxidized and reduced glutathione in fresh controls (=CF), cut material (=C), cut and 5 min boiled material (=CB5) and cut and 20 min boiled material (=CB20). Percentage was calculated relating to total glutathione amounts. Same letters mean no significant difference. P < 0.05 analyzed by Kruskal-Wallis-ANOVA, n = 10.

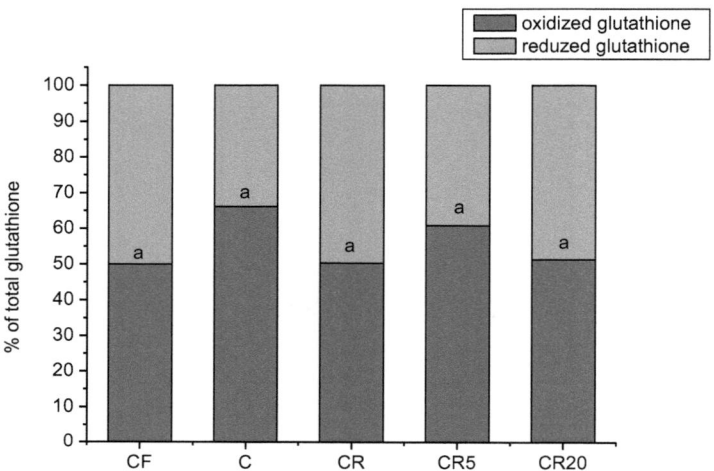

Figure 86: Percentage of oxidized and reduced glutathione in fresh controls (=CF), cut material (=C), cut and let rested material (=CR), cut, let rested material boiled for 5 min (=CR5) and for 20 min (=CR20). Percentage was calculated relating to total glutathione amounts. Same letters mean no significant difference. P < 0.05 analyzed by Kruskal-Wallis-ANOVA, n = 10.

Calculated in % of fresh controls dried material showed an increase of about 100%, dried and 5 as well as 20 minutes boiled material an decrease of -70 to -80% of total glutathione contents (*Fig. 87*). An decrease of glutathione content was found in cut material (about -5%), in cut and 5 respectively 20 minutes boiled material (about -90% and -85%) (*Fig. 88*). Glutathione contents in cut as well as let rested material increased by +10%, cut material (-5%) and cut, let rested and 5 minutes as well as 20 minutes boiled material decreased by about -90% (*Fig. 89*).

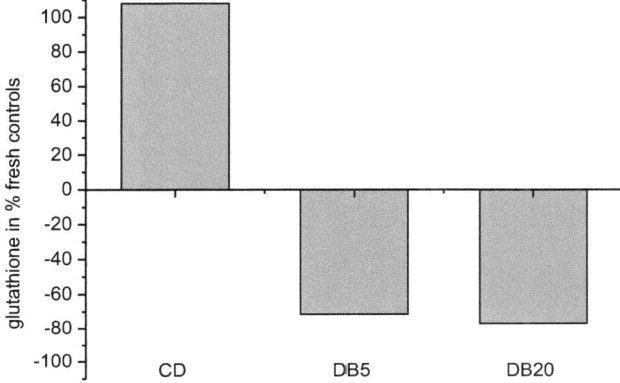

Figure 87: Increase/decrease of glutathione contents in dried controls (=CD), dried and 5 min boiled (=DB5) and dried and 20 min boiled (=DB20) calculated in % of fresh controls. n = 10.

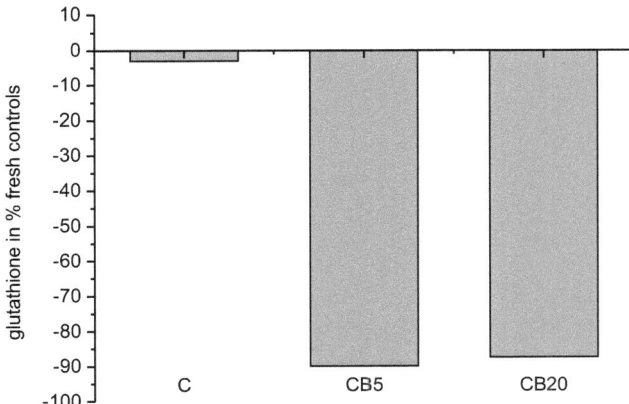

Figure 88: Decrease of ascorbate in cut material (=C), cut and 5 min boiled material (=CB5) and cut and 20 min boiled material (=CB20) calculated in % of fresh controls. n = 10.

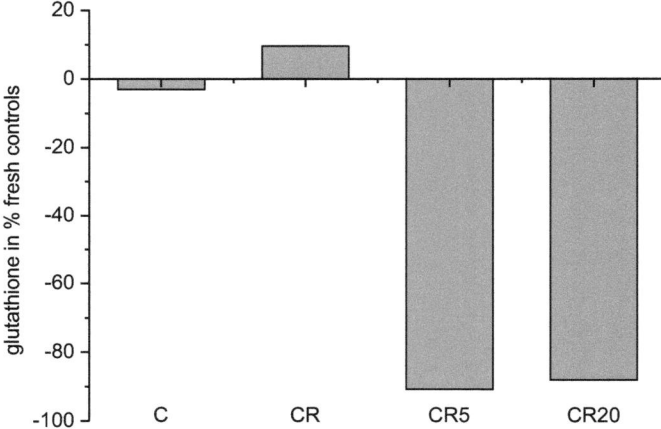

Figure 89: Increase/decrease of glutathione in cut material (=C), cut and let rested material (=CR), cut, let rested material boiled for 5 min (=CR5) and for 20 min (=CR20) calculated in % of fresh controls. n = 10.

6.1.5 Summary of the results of nasturtium

Table 11 shows the significant changes of all examined substances in dried and boiled material (5 min and 20 min) compared to dried controls as an overview and summary.

Table 11: Summary of significant changes in dried and 5 and 20 minutes boiled nasturtium material compared to dried controls. ↑ = increase, ↓ = decrease, ↔ = no significant change, black arrows: compared to control material, red arrows in brackets: compared to 5 minutes boiled material.

Nasturtium	dried	
	5 min	20 min
chlorophylls	↑	↑ (↓)
carotenoids	↑	↑ (↓)
chlorophyll a/b	↔	↔ (↔)
α-tocopherol	↓	↔ (↑)
ascorbate	↓	↓ (↑)
glutathione	↓	↓ (↔)

Table 12 shows the significant changes of all examined substances in fresh cut and boiled material (5 min and 20 min) compared to fresh controls as well as let rested material (cut, 5 min and 20 min boiled) as an overview and summary.

Table 12: Summary of significant changes in fresh cut and 5 and 20 minutes boiled nasturtium material as well as let rested material (cut, 5 and 20 min boiled) compared to fresh controls. ↑ = increase, ↓ = decrease, ↔ = no significant change, black arrows: compared to control material, red arrows in brackets: compared to 5 minutes boiled material.

Nasturtium	fresh				fresh and rested			
	cut	5 min	20 min		cut	5 min	20 min	
chlorophylls	↑	↑	↑	(↓)	↑	↑	↔	(↓)
carotenoids	↑	↑	↑	(↓)	↑	↑	↑	(↓)
chlorophyll a/b	↔	↔	↔	(↔)	↔	↔	↔	(↔)
α-tocopherol	↔	↑	↑	(↔)	↔	↑	↑	(↔)
ascorbate	↔	↓	↓	(↑)	↔	↓	↓	(↔)
glutathione	↔	↓	↓	(↔)	↔	↓	↓	(↔)

6.2 Discussion of the processing experiments of nasturtium

In general, it is well known that various factors such as temperature and pH of the media, processing treatment (e.g. heat) and storage can have a strong influence on the activity and concentrations of antioxidants (GAZZANI & al. 1998). Especially in fruits and vegetables the antioxidative capacity can be affected by cultivars, maturity and environmental conditions e.g. sunlight exposure as well as by preparation, processing and cooking (PRIOR & CAO 2000). But not only fruits, also spices or seasoning herbs are natural sources of antioxidants. Furthermore, they are often used as supplements for food (both, industrially and at home) affecting flavor and enhancing food quality due to their antioxidants which can prevent or delay oxidative deterioration during processing and storage (COSIO & al. 2006, AL-DUAIS & al. 2009). Therefore, it is necessary to examine changes in contents of herbs during culinary processing, too. In industrial food processing thermal treatment holds an important place, because of its many benefits for food preservation. But it has both, negative and positive impact on antioxidants. The positive effects include inactivation of oxidases and breakdown of food structures leading to improved bioavailability, the negative effect is the degradation of various antioxidative compounds (e.g. carotenoids and ascorbic acid) due to their sensitivity to heating (DIPLOCK & al. 1998). Blanching is one of the pre-treatments used before freezing, inactivating enzymes and preventing biochemical reactions (e.g. discoloration) in the frozen product (MOUNTNEY & GOULD 1988). However, blanching causes undesirable changes in the food properties which may result in the loss of color, flavor, texture and nutrients (PALA 1983, PIZZOCARO & al. 1995). It is known that blanching decreases the levels of water-soluble constituents. On the other hand, the thermal inactivation of enzymes limits the degradation of both, chlorophylls and carotenoids (HEATON & al. 1996). Another positive effect of blanching is the de-aeration and in series, the reduced content of oxygen in plant tissue leads to a better preservation of pigments (TOIVONEN 1997). Especially, the type of cooking can not only affect contents of various substances, but also the total antioxidant capacity. LAKO & al. (2007), for example, showed that steamed *Moringa oleifera* leaves had a higher total antioxidant capacity than the boiled ones (CASTENMILLER & al. 1999).

In our studies we simulated processing and cooking at home. Therefore, dried plant material as well as fresh leaves of nasturtium were processed in different ways. Dried material was boiled for 5 and 20 minutes, fresh material was cut and immediately boiled for 5 and 20 minutes or first cut, let rested and then boiled for 5 and 20 minutes. In series, contents of chlorophylls, carotenoids, α-tocopherol, ascorbate and glutathione were examined.

Degradation of chlorophylls during thermal processes is the main cause of green color loss (which is a problem for food industry) (ESKIN 1990) and also carotenoids are often victim of degradation due to their highly unsaturated structure responsible for their sensitivity to heat, oxygen and light (RODRIGUEZ-AMAYA 2002, reviewed in LEŠKOVÁ & al. 2006; RODRIGUEZ & RODRIGUEZ-AMAYA 2007). Interestingly, our studies showed an increase of pigments (chlorophylls and carotenoids) after all processing variants compared to controls (expressed on dry matter basis). LISIEWSKA & al. (2004) showed that blanching of dill did not affect the level of carotenoids if the results were expressed on fresh matter basis, but after calculating the results on a dry matter basis, the level of carotenoids increased after this treatment. Also KHACHIK & al. (1992) stressed that, in leafy vegetables (e.g. spinach), the content of carotenoids increased after the thermal treatment. A possible explanation for the determined greater content after thermal treatment was assumed by GRANADO & al. (1992), who attribute this effect to the denaturation of carotenoid complexes, permitting a complete extraction of these compounds. On the other hand, MOSHA & al. (1997) did not only show an increase of carotenoid content in four species, but also a decrease in two others. GONÇALVES & al. (2009) found out, that colour parameters significantly changed in blanched watercress (*Nasturtium officinale*), the latter became lighter and more green and yellow. Furthermore, total chlorophylls decreased significantly, whereas chlorophyll a decreased faster than chlorophyll b. Similar results were found in peas (GÖKMEN & al. 2005). These results are oppositional to our increase which could be due to different calculations (their results were expressed in fresh matter, cp. LISIEWSKA & al. 2004) or different processing methods. GONÇALVES & al. (2009) blanched the water cress samples for only 20 seconds, we boiled our material for much longer (5 respectively 20 minutes). However, blanching of dill plants did not affect significant changes in the content of chlorophylls (LISIEWSKA & al. 2004) and various other authors examined a loss of chlorophylls in blanched leafy vegetables (LISIEWSKA & KMIECIK 1997, SONG & al. 2003). Altogether, pigment content variations depend on various factors, e.g. not only the pH of the used water (GUNAVAN & BARRINGER 2000), but also the temperature seems to be important for changes during processing. An increase in blanching temperature by 10°C increased the reaction rate constant of greenness in vegetable soybeans by a factor of two, and this indicated an Arrhenius relationship (describes the temperature dependence of chemical reactions) (SONG & al. 2003). Furthermore, differences in content changes are found between stir-frying, blanching and boiling In boiling, water is absorbed by the food, causing dilution of the carotenoids, whereas stir-frying (because of water loss) may concentrate the carotenoids, giving higher carotenoid content per unit weight of vegetable (DE SA & RODRIGUEZ-AMAYA 2003). During boiling, leaching of phytochemicals into the cooking liquid occurs (CASTENMILLER & al. 1999) which is not so high during blanching (because of the shorter time). In general, not only the changes of total pigment contents are important, but also variations in their composition. DE SA & RODRIGUEZ-AMAYA (2003) assumed that cooked vegetables would have variations in their carotenoid composition not only due to varying cooking conditions, but also compositional differences of the raw material (e.g. stage of maturity, cultivar, part of the plant utilized, climatic or seasonal effects, agricultural and post-harvest handling) (DE SA & RODRIGUEZ-AMAYA 2003). Besides processing conditions also different classes of pigments differ in their stability towards heat treatment (KHACKIK & al. 1991, GONÇALVES & al. 2009). For example, violaxanthin was reported as very labile during heat treatment (KHACHIK & al. 1992), β-carotene showed an increase during cooking/steaming (KHACKIK & al. 1991, LAKO & al. 2007) and chlorophyll a decreased faster than chlorophyll b during blanching (GONÇALVES & al. 2009).

Ascorbate (vitamin C) is well examined, because of its importance to human health and as one of the major antioxidants in plant species (see e.g. SMIRNOFF 2000). However, water-soluble antioxidants such as ascorbic acid are also sensitive to heating and therefore, thermal treatment can induce significant changes not only in chlorophylls and carotenoids, but also in vitamin C contents (GONÇALVES & al. 2009). In general, various studies (e.g. NEGI & ROY 2004, DAOOD & al. 2006) showed that the degradation of ascorbic acid (due to processing and storage) is depending on temperature and duration following first order kinetics (GUTZEIT & al. 2008). In our studies only the cutting of material showed no influence on ascorbic acid content in the plant material (by contrast, UCKIAH & al. (2009) showed that during peeling of fresh pineapple vitamin C degraded). On the other hand, after all boiling experiments the content of ascorbic acid was significantly decreased (total contents between -80 and -100% compared to controls!). Actually, various studies confirm our results showing degradation of ascorbic acid because of thermal treatment. For example, after boiling various green vegetables (e.g. broccoli, cabbage, collard, mustard green, spinach and Swiss chard) showed a significant decrease in ascorbic acid contents (VANDERSLICE & al. 1990, GIL & al. 1998, GIL & al. 1999). Cooking of African leafy vegetables (*Cassia tora, Corchorus tridens*) resulted in dramatic losses, too (PRABHU & BARRETT 2009) and also the more gentle cooking variant "blanching" for 1 minute showed similar results in parsley (LISIEWSKA & KMIECIK 1997) and vegetable soybeans (SONG & al. 2003). Blanching for only 20 seconds led to a decrease in total vitamin C contents in water cress (*Nasturtium officinale*) from about 65 mg/100g fresh weight (controls) to about 11 mg/100g fresh weight (CRUZ & al. 2009) and also GONÇALVES & al. (2009) showed by trend (although not significant) a decreased ascorbic acid content in water cress after blanching. GARROTE & al. (1986) assumed, that the main mechanisms of ascorbic acid losses during blanching are thermal induced degradation or leaching (GONÇALVES & al. 2009). However, for instance, during preparation of teas leaching is requested enriching it with minerals (ÖZCAN & al. 2008). In our studies, not only ascorbic acid, but also glutathione contents showed a decrease after boiling. These results are confirmed by ZIELIŃSKY & al. (2006) and FERNANDEZ-OROZCO & al. (2003) who showed also losses of glutathione in cooked buckwheat groats and lentil cultivars. Because of the water solubility of glutathione we assume, that the examined losses could be (similar to ascorbic acid) due to leaching.

Tocopherols are highly lipophilic and operative in membranes (e.g. cell membrane, intra- and extracellulary (RODRIGO & al.2007) and lipoproteins and therefore, their most important antioxidant function appears to be the inhibition of lipid peroxidation (DIPLOCK & al. 1998). To increase food quality and prevent or delay oxidative deterioration (e.g. during processing and storage) (COSIO & al. 2006), tocopherol is used as food additive (DIPLOCK 1992, LANGSETH 1995). In general, (similar to other antioxidants) the degradation of tocopherols seems to be depending on temperature and (storage) duration following first order kinetics, too (DAOOD & al. 2006, LAVELLI & al. 2006, HIDALGO & al. 2009, SABLIOV & al. 2009). In processing of hot and sweet pepper for example, vitamin E was more decreased the higher the temperature and pressure was (PERUCKA & MATERSKA 2007). In our studies α-tocopherol content is increased after the boiling experiments of fresh material (contents nearly doubled!) which is similar to our results for (also lipid soluble) pigments. AL-DUAIS & al. (2009) also showed enhanced vitamin E contents in processed (30 minutes boiled and in series dried) *Cyphostemma digitatum* leaves and attributed this fact to two reasons: The processing may cause denaturation of proteins and a complete destruction of cell walls and organelles which consequently results in liberation of vitamin E from the lipids becoming more available for extraction. Similar reasons were given by GRANADO & al. (1992) for increased carotenoic contents. On the other hand, extrusion cooking of buckwheat groats (120 – 200°C) caused a significant decrease (about 63%) in vitamin E content (ZIELINSKY & al. 2006) and also cooking of lentils (FERNANDEZ-OROZCO & al. 2003) and heating of olive oil (BRENES & al. 2002) led to losses in tocopherol contents. These results let us assume, that various factors like e.g. (plant) material, temperature and processing time are important for modifications in tocopherol contents which complicate comparisons.

In conclusion, preparation, processing and cooking can cause changes in both, antioxidant contents and capacity (e.g. reviewed in MADSEN & BERTELSEN 1995, PRIOR & CAO 2000, CHOHAN & al. 2008). These changes, mainly losses, vary widely according to cooking method and type of (plant) food and the degradation depends on specific parameters during the culinary process, e.g. temperature, oxygen, light, moisture, pH, and length of exposure (reviewed in LEŠKOVÁ & al. 2006). Besides losses, on the other hand it should be noted, that some phytochemicals may be more bio-available when the food is cooked (ROCK & al. 1998). For example, carotenoid bioavailability could be enhanced by heat treatment (HOF & al. 2000) and ascorbic acid was more bio-available from cooked broccoli than it was from the raw form (reviewed in DIPLOCK & al. 1998). Another benefit of processed samples is improved aroma, because many aroma compounds are generated from carotenoids by food cooking and processing (LEFFINGWELL 2002). Especially for food industry, (to increase bioavailability and dietary intake), it is recommended to optimize cooking and processing conditions (AL-DUAIS & al. 2009) and to test spices and herbs accordingly in the actual food under realistic conditions prior practical use (YANISHLIEVA & al. 2006). In general, even if antioxidants are non-nutrient they may contribute to the total antioxidant effects of the diet in vivo as well as during processing, too (DIPLOCK & al. 1998) and the variability e.g. in the compositions of antioxidants in different (plant) foods indicate the importance of eating a variety of food sources (LAKO & al. 2007) in both, raw and processed form.

7. Conclusions

As the results of this work and various others studies show, the responses of plants to stress as well as the effect of processing and storage on antioxidants are depending on numerous factors. However, the aim was to answer some questions which is done now.

It is well known, that **drought** affects the plant metabolism, concentrations of substances and enzyme activities are changed and photosynthesis can be influenced. In series, metabolic adaptations can lead to faster and/or stronger reactions to further stress attacks. *How do stress and recovery influence the metabolism of seasoning herbs?* The results show not only that there is an influence but also how drought and recovery affects photosynthesis and the concentration of antioxidants in the seasoning herbs nasturtium, borage and summer savory. It can be seen clearly, that stress responses vary between plant species and are related to the duration and intensity of stress. *Are there differences between the plant species?* Definitely yes! *Are plants "memorizing" the stress by adapting their metabolism?* Changes of some substances and photosynthetic parameters and especially the differences in the light curves of the re-watered plants seem to be indicative. *Can mild drought stress be used to prime the plants to be better prepared for further stress attacks?* In case of nasturtium the higher effectiveness of photosynthesis and the higher tocopherol content of the re-watered nasturtium plants express such advantage. For borage and summer savory the results proved to be not consistent so that further studies would be necessary (e.g. light curves and photosynthetic parameters, other stress intensities…). *Are the examined plant species good antioxidant sources?* Comparing pigment, tocopherol, ascorbate and glutathione of all 3 plant species, nasturtium has the highest (~double compared to borage and sarvory!) contents. *Is mild drought stress a possibility to enhance antioxidant contents in plant material for e.g. (food) industry?* For tocopherol in nasturtium this would definitively be an option.

Many studies show that antioxidants are sensitive to e.g. water loss, light or heat (according to their chemical properties) and can therefore degrade easily during **drying and storage**. *Are the antioxidants degrading during drying and storage in plant material - if yes, how massive? Do differences between antioxidants exist?* There are (mostly) degradations in the contents of antioxidants whereas especially the water-soluble substances ascorbate and glutathione are strongly affected and sometimes it is even not possible to detect them anymore (exception: increase of tocopherol and glutathione in nasturtium). *Are there even antioxidants left after drying and storage for months?* Lipid-soluble substances (pigments and tocopherol) are left; water-soluble ascorbate not (exception: nasturtium) and glutathione is (mostly) left, although in low amounts (exception: nasturtium). *Is there a difference between storage in paper bags and tie bags?* The differences are low but mostly significant. Surprisingly, the simple paper bags showed better results compared to tie bags (which are used for plants containing essential oils). *Is there a difference between the plant species?* Altogether, there is a trend, showing degradation of substances during drying and storage in all three plant species, but the changes are species-dependent. *Is one of the examined species a good antioxidant source even after drying and storage and if yes, for what antioxidants?* Dried and stored plant material of nasturtium has good amounts of all examined substances (exception: ascorbate in stored material) and borage and summer savory can be at least good sources for the lipid-soluble antioxidants.

Due to their chemical properties antioxidants are not only sensitive to drying and storage but also to different **processing** methods which are quite common in industrial food production. *Does the processing of the plant material lead to a loss of antioxidants and if yes, how massive?* In this study and in various others increases, decreases or no changes are found, depending on plant species, substances as well as processing methods and duration. *Is nasturtium a good antioxidant source even after processing and if yes, for what antioxidants? Can antioxidants even be found in dried and processed plant material?*! Processed nasturtium can be a good source for lipid-soluble antioxidants and due to processing the bioavailability of the substances in the plant material can even be enhanced. *Is it advisable to add (fresh or dried) herbs as antioxidant source to food?* Nasturtium, borage and summer savory can be good sources for lipid-soluble antioxidants. Furthermore, adding herbs can also enhance flavour and they contain various other substances which contribute to the total antioxidant capacity of food. *How does the processing method or duration influence the degradation of antioxidants?* In general, a trend can be seen showing that degradations of antioxidants follow (often) first order kinetics, i.e. the higher the processing temperature / the longer the processing duration, the higher the degradation of antioxidants and vice versa. But it has to be kept in mind, that antioxidants are often part of e.g. protein complexes which degrade during processing first and therefore 1) protect the antioxidants and 2) enhance their bioavailability. *In which point of time is it advisable to add herbs (as antioxidant sources)?* Due to the chemical properties of antioxidants and their degradation (mentioned above) adding herbs is advisable at the end of processing, keeping duration (and potential degradation) short.

In conclusion, this work shows that the content of antioxidants in plants (during stress) and their degradation (during drying, storage and processing) are related to environmental factors during growth, harvesting conditions and post-processing.

8. References

AL-DUAIS M., HOHBEIN J., WERNER S., BÖHM V. & JETSCHKE G. 2009. Contents of vitamin C, carotenoids, tocopherols and tocotrienols in the subtropical plant species *Cyphostemma digitatum* as affected by processing. - Food Chem. 57: 5420-5427.

AN J., ZHANG M., LU Q. & ZHANG Z. 2006. Effect of a prestorage treatment with 6-benzylaminopurine and modified atmosphere packaging storage on the respiration and quality of green asparagus spears. – J. Food Engin. 77(4): 951–957.

APEL K. & HIRT H. 2004. Reactive Oxygen Species: Metabolism, Oxidative Stress, and Signal Transduction – Review. - Annu. Rev. Plant Biol. 55: 373–399.

ARSLAN D. & ÖZCAN M. M. 2010. Study the effect of sun, oven and microwave drying in quality of onion slices. - LWT - Food Sci. Tech. 43: 1121-1127.

ASADA K. 1999. The water-water cycle in chloroplasts: scavenging of active oxygens and dissipation of excess photons. - Annu Rev. Plant Physiol. Plant Mol. Biol. 50: 601–639.

ASADA K. 2006. Production and scavenging of reactive oxygen species in chloroplasts and their functions. - Plant Physiol. 141: 391–396.

ASADA K. & TAKAHASHI M. 1987. Production and scavenging of active oxygen in chloroplasts. – In: KYLE D. J., OSMOND C. B. & ARNTZEN C. J. (eds.), Photoinhibition, pp227–287, Elsevier. Amsterdam.

ASHRAF M. 2003. Relationships between leaf gas exchange characteristics and growth of differently adapted populations of blue panic grass (*Panicum antidotale* Retz.) under salinity or waterlogging. - Plant Sci. 165: 69–75.

BAHER Z.F., MIRZA M., GHORBANLI M. & REZAII M. B. 2002. The influence of water stress on plant height, herbal and essential oil yield and composition in *Satureja hortensis* L. - Flavour Fragr.J. 17: 275–277.

BARR D. E. 2001. Potential of evening primrose, borage, black-currant, and fungal oils in human health. - Ann Nutr Metab. 45: 47–57.

BARTH C., MOEDER W., KLESSIG D. F. & CONKLIN P. L. 2004. The Timing of Senescence and Response to Pathogens Is Altered in the Ascorbate-Deficient *Arabidopsis* Mutant *vitamin c-1*. - Plant Physiol. 134: 1784–1792.

BARTOLI C. G., SIMONTACCHI M., TAMBUSSI E., BELTRANO J., MONTALDI E. & PUNTARULO S. 1999. Drought and watering-dependent oxidative stress: effect on antioxidant content in *Triticum aestivum* L. leaves. - J. Exp. Bot. 50: 373-381.

BAYDAR H., SAGDIC O., OZKAN G. & KARADOGAN T. 2004. Antibacterial activity and composition of essential oils from *Origanum, Thymbra* and *Satureja* species with commercial importance in Turkey. - Food Control 15: 169–172.

BEAUBAIRE N.A. & SIMON J.E. 1987. Production potential of borage (*Borago officinalis* L.). - Acta Hort. 208:101-114.

BEUCHAT L. R. 1994. Antimicrobial properties of spices and their essential oils. - In: DILLON V. M. & BOARD R. G. (Eds.), Natural Antimicrobials Systems and Food Preservation, pp. 167-180. CAB International, Wallingford. UK.

BEZIĆ N., ŠAMANIĆ I., DUNKIĆ V., BESENDORFER V. & PUIZINA J. 2009. Essential Oil Composition and Internal Transcribed Spacer (ITS) Sequence Variability of Four South-Croatian Satureja Species (Lamiaceae). - Molecules 14: 925-938.

BICKERICH G., DÖRFLER H. P., HILLER K. & ROSELT G. 2001. Das große Hausbuch der Arznei-, Heil- und Giftpflanzen. – Urania Verlag. Berlin.

BINET L. 1964. A biologist physician in the country. - Medical Biology (Paris) 53: 5–28.

BLOEM E., HANEKLAUS S. & SCHNUG E. 2007. Comparative effects of sulfur and nitrogen fertilization and post-harvest processing parameters on the glucotropaeolin content of *Tropaeolum majus* L. – J. Sci. Food Agric. 87: 1576-1585.

BLOKHINA O., VIROLAINEN E. & FAGERSTEDT K. V. 2003. Antioxidants, oxidative damage and oxygen deprivation stress: a review. – Ann. of Bot. 91: 179-194.

BOELT B. 1990. Seed rate, sowing time and harvest time in dill (*Anethum graveolens* L.) for freeze drying. - Tidsskrift for Planteavl 94: 497–502.

BOO Y. C. & JUNG J. 1999. Waterdeficit-induced oxidative stress and antioxidative defences in rice plants. - J. Plant Physiol. 155: 255–261.

BOTELLA-PAVÍA P. & RODRÍGUEZ-CONCEPCIÓN M. 2006. Carotenoid biotechnology in plants for nutritionally improved foods. Review. – Physiol. Plant. 126: 369–381.

BOYRAZ N. & ÖZCAN M. 2006. Inhibition of phytopathogenic fungi by essential oil, hydrosol, ground material and extract of summer savory (*Satureja hortensis* L.) growing wild in Turkey. - Inter. J. Food Microbiol. 107: 238-242.

BOZIN B., MIMICA-DUKIC N., SAMOJLIK I. & JOVIN E 2007. Antimicrobial and antioxidant properties of rosemary and sage (*Rosmarinus officinalis* L. and *Salvia officinalis* L., Lamiaceae) essential oils. - J. Agric. Food Chem. 55: 7879-7885.

BREME K., FERNANDEZ X., MEIERHENRICH U. J., BREVARD H. & JOULAIN D. 2007. Identification of new, odor-active thiocarbamates in cress extracts and structure-activity studies on synthesized homologues. - J.Agric. Food Chem. 55: 1932–1938.

BREME K., GUILLAMON N., FERNANDEZ X., TOURNAYRE P., BREVARD H., JOULAIN D., BERDAGUE J. L. & MEIERHENRICH U. J. 2009. First Identification of O,S-Diethyl Thiocarbonate in Indian Cress Absolute and Odor Evaluation of Its Synthesized Homologues by GC-Sniffing. - J. Agric. Food Chem. 57: 2503–2507.

BRENES M., GARCÍA A., DOBARGANES M. C., VELASCO J. & ROMERO C. 2002. Influence of Thermal Treatments Simulating Cooking Processes on the Polyphenol Content in Virgin Olive Oil. - J.Agric. Food Chem. 50: 5962–5967.

BRUCE T. J. A., MATTHES M. C., NAPIER J. A. & PICKETT J. A. 2007. Stressful "memories" of plants: Evidence and possible mechanisms. Review. - Plant Sci. 173: 603-608.

BUCHAILLOT A., CAFFIN N. & BHANDARI B. 2009. Drying of Lemon Myrtle (*Backhousia citriodora*) leaves: retention of volatiles and color. - Drying Technology 27(3): 445-450.

BURDOCK G. A. (ed.) 1995. - Fenaroli's Handbook of Flavor Ingredients, 3rd edition, vol. 1. CRC Press, Boca Raton, Florida.

BURTON G. W. & INGOLD K. U. 1984. β-Carotene: an unusual type of lipid antioxidant. - Science 224: 569-573.

CAI Z.-Q., CHEN Y.-J., GUO Y.-H. & CAO K.-F. 2005. Responses of two field-grown coffee species to drought and re-hydration. - Photosynthetica 43: 187-193.

CAILLET S., YU H., LESSARD S., LAMOUREUX G., AJDUKOVIC D. & LACROIX M. 2007. Fenton reaction applied for screening natural antioxidants. - Food Chem. 100: 542–552.

CAPECKA E., MARECZEK A. & LEJA M. 2005. Antioxidant activity of fresh and dry herbs of some Lamiaceae species. - Food Chem. 93: 223–226.

CASTENMILLER J. J. M., WEST C. E., LINSEEN J. P. H., HOF K. H. & VORAGEN A. G. J. 1999. The food matrix of spinach is a limiting factor in determining the bioavailability of β-carotene and to a lesser extent of lutein in humans. – J. Nutr. 129: 349–355.

CAZZONELLI C. I. & POGSON B. J. 2010. Source to sink: regulation of carotenoid biosynthesis in plants. Review. - Trends Plant Sci. 15(5): 266-274.

CEBALLOS C. & FERNÁNDEZ H. 2000. Synthetic antioxidants determination in lard and vegetable oils by the use of voltammetric methods on disk ultramicroelectrodes. - Food Res. Inter. 33(5): 357-365.

CHAVES M.M., MAROCO J. P. & PEREIRA J. S. 2003. Understanding plant responses to drought - from genes to whole plant. - Funct. Plant Biol. 30: 239-264.

CHAVES M. M. & OLIVEIRA M. M. 2004. Mechanisms underlying plant resilience to water deficits: prospects for water-saving agriculture. - J. Exp. Bot. 55: 2365-2384.

CHEVALLIER A. 2000. Die BLV Enzyklopädie der Heilpflanzen. - BLV Verlagsgesellschaft mbH. München.

CHOHAN M., FORSTER-WILKINS G. & OPARA E. I. 2008. Determination of the Antioxidant Capacity of Culinary Herbs Subjected to Various Cooking and Storage Processes Using the ABTS*+ Radical Cation Assay. - Plant Foods Hum. Nutr. 63: 47–52.

CONNER D.E. & BEUCHAT L. R. 1984. Effects of essential oils from plants on growth of food spoilage yeasts. - J. Food Sci. 49: 429-434.

CONNER D.E. 1993. Naturally occurring compounds. - In: DAVIDSON P. M. & BRANEN A. L. (eds.), Antimicrobials in Foods. Marcel Dekker. New York.

CORRÊA M. P. 1978. Dicionário das plantas úteis do Brasil e das exóticas cultivadas, vol. 6. - Imprensa Nacional. Rio de Janeiro.

COSIO M. S., BURATTI S., MANNINO S. & BENEDETTI S. 2006. Use of an electrochemical method to evaluate the antioxidant activity of herb extracts from the Labiatae family. - Food Chem. 97: 725-731.

COSTA M. L., CIVELLO P. M., CHAVES A. R. & MARTÍNEZ G. A. 2005. Effect of hot air treatments on senescence and quality parameters of harvested broccoli (*Brassica oleracea* L. var *italica*) heads. – J. Sci. Food Agric. 85(7): 1154–1160.

CROWELL P.L. 1999. Prevention therapy of cancer by dietary monoterpenes. – J. Nutr. 129: 775-778.

References

CRUZ R.M.S., VIEIRA M. C. & SILVA C. L. M. 2009. Effect of cold chain temperature abuses on the quality of frozen watercress (*Nasturtium officinale* R.Br.). – J. Food Engin.94: 90–97.

CRUZ DE CARVALHO M. H. 2008. Drought stress and reactive oxygen species: production, scavening and signalling. - Plant Signal Behav. 3: 156–165.

CRUZ DE CARVALHO M. H., BRUNET J., BAZIN J., KRANNER I., D'ARCY-LAMETA A., ZUILY-FODIL Y. & CONTOUR-ANSEL D. 2010. Homoglutathione synthetase and glutathione synthetase in drought-stressed cowpea leaves: expression patterns and accumulation of low-molecular-weight thiols. – J. Plant Physiol. 167(6): 480-487.

CUI K., LUO X. L., XU K. Y. & MURTHY M. R. V. 2004. Role of oxidative stress in neuro-degeneration: recent developments in assay methods for oxidative stress and nutraceutical antioxidants. - Prog. Neuropsychopharmacol. Biol. Psychiatry 28: 771–799.

DACHLER M. & PELZMANN H. 1999. Arznei- und Gewürzpflanzen: Anbau, Ernte, Aufbereitung. - Österreichischer Agrarverlag, Klosterneuburg.

DAOOD H. G., KAPITÁNY J., BIACS P. & ALBRECHT K. 2006. Drying temperature, endogenous antioxidants and capsaicinoids affect carotenoid stability in paprika (red pepper spice). – J. Sci. Food Agric. 86: 2450–2457.

DAT J., VANDENABEELE S., VRANOVÁ E., VAN MONTAGU M., INZÉ D. & VAN BREUSEGEM F. 2000. Dual action of the active oxygen species during plant stress responses. - Cell Mol. Life Sci. 57: 779-795.

DAVEY M.W., VAN MONTAGU M., INZÉ D., SANMARTIN M., KANELLIS A., SMIRNOFF N., BENZIE I. J. J., STRAIN J. J., FAVELL D. & FLETCHER J. 2000. Plant L-ascorbic acid: chemistry, function, metabolism, bioavailability and effects of processing – Review. – J. Sci. Food Agric. 80: 825-860.

DAVIES B. H. 1976. Carotenoids. – In: GOODWIN T. W. (ed.), Chemistry and Biochemistry of Plant Pigments, 2nd edition, vol. 2. Academic Press. London, NewYork, San Francisco.

DE MEDEIROS J. M. R., MACEDO M., CONTANCIA J. P., NGUYEN C., CUNNINGHAM G. & MILES D.H. 2000. Antithrombin activity of medicinal plants of the Azores. – J. Ethnopharm. 72: 157–165.

DE OLIVEIRA G. P. R. & RODRIGUEZ-AMAYA D. B. 2007. Processed and Prepared Corn Products As Sources of Lutein and Zeaxanthin: Compositional Variation in the Food Chain. J. Food Sci. 72(1): 79-85.

DE SA M. C. & RODRIGUEZ-AMAYA D. B. 2003. Carotenoid composition of cooked green vegetables from restaurants. - Food Chem. 83: 595–600.

DEANS S. G. & RITCHIE G. 1987. Antibacterial properties of plant essential oils. - Int. J. Food Microbiol. 5: 165-180.

DEANS S.G. & SVOBODA K. P. 1989. Antibacterial activity of summer savory (*Satureja hortensis* L.) essential oil and its constituents. - J. Hort. Sci. 64: 205–211.

DELLAPENNA D. & LAST R. L. 2006. Progress in the dissection and manipulation of plant vitamin E biosynthesis – Review. – Physiol. Plant. 126: 356–368.

DEMMING-ADAMS B. & ADAMS W. W. 1992. Photoprotection and other responses of plants to high light stress. - Ann. Rev. Plant Physiol. Plant Mol. Biol 43: 599-626.

DERIDDER B. P. & GOLDSBROUGH P. B. 2006. Organ-specific expression of glutathione S-transferases and the efficacy of herbicide safeners in *Arabidopsis*. - Plant Physiol. 140: 167–175.

References

DEROSSI A., DEPILLI T. & FIORE A. G. 2010. Vitamin C kinetic degradation of strawberry juice stored under non-isothermal conditions. - LWT - Food Sci. Tech. 43: 590–595.

DESIKAN R., MACKERNESS S. A. H., HANCOCK J. T. & NEILL S. 2001. Regulation of the *Arabidopsis* transcriptome by oxidative stress. - Plant Physiol. 127: 159–172.

DESMET P.A.G.M. 1991. The safety of borage oil seed. – Can. Pharm. J. 124: 5.

DIPLOCK A. T. 1992. The Role of antioxidant nutrients in disease. - Health Nutr. Inform. 3: 1214-1217.

DIPLOCK A. T., CHARLEUX J. L., CROZIER-WILLI G., KOK F. J., RICE-EVANS C., ROBERFROID M., STAHL W. & VINA-RIBES J. 1998. Functional food science and defence against reactive oxidative species. - Brit. J. Nutr. 80: 77–112.

DRAGLAND S., SENOO H., WAKE K., HOLTE K. & BLOMHOFF R. 2003. Several culinary and medicinal herbs are important sources of dietary antioxidants. – J. Nutr. 133: 1286–1290.

DZIEZAK J. D. 1986. Preservatives: antimicrobial agents. A means toward product stability. – Food Tech. 40(9): 104–111.

EDWARDS R., BRAZIER-HICKS M., DIXON D. P. & CUMMINS I. 2005. Chemical manipulation of antioxidant defences in plants. – Adv. Bot. Res. 42: 1–32.

EGERT M. & TEVINI M. 2002. Influence of drought on some physiological parameters symptomatic for oxidative stress in leaves of chives (*Allium schoenoprasum*). - Env. Exp. Bot. 48: 43–49.

ESKIN N. A. M. 1990. Biochemical changes in raw food: fruit and vegetables; V-color changes. – In: Biochemistry of food, 2nd edition, pp. 90-119. - Academic Press INC. SanDiego, CA.

ESKIN N. A. M. & TAMIR S. 2006. Dictionary of Nutraceuticals and Functional Foods, pp. 176–178. - CRC & Taylor & Francis. Boca Raton, USA.

FALK J. & MUNNE-BOSCH S. 2010. Tocochromanol functions in plants: antioxidation and beyond. Darwin Review. – J. of Exp. Bot. 61(6): 1549-1566.

FARRÉ G., SANAHUJA G., NAQVI S., BAI C., CAPELL T., ZHU C. & CHRISTOU P. 2010. Travel advice on the road to carotenoids in plants - Review. - Plant Sci. doi: 10.1016/j.plantsci.2010.03.009.

FERNANDEZ-OROZCO R., ZIELIŃSKI H. & PISKUŁA M. K. 2003. Contribution of low-molecular-weight antioxidants to the antioxidant capacity of raw and processed lentil seeds. - Food 47(5): 291–299.

FERREIRA R. B. G., VIEIRA M. C. & ZÁRETE N. A. H. 2004. Análise de crescimento de *Tropaeolum majus* 'jewel' em função de espaçamentos entre plantas. – Rev. Bras. Pl. Med. 7(1): 57–66.

FERRO D. 2006. Fitoterapia: conceitos clínicos, 410 pp. - Atheneu. SãoPaulo.

FLEXAS J., BOTA J., GALMÉS J., MEDRANO H. & RIBAS-CARBO M. 2006. Keeping a positive carbon balance under adverse conditions: responses of photosynthesis and respiration to water stress. - Physiol. Plant. 127: 343–352.

FOYER C. H. & NOCTOR G. 2009. Redox regulation in photosynthetic organisms: signalling, acclimation, and practical implications. Review. - Antioxid. Redox. Signaling 11: 861-905.

FOYER C. H., THEODOULOU F.L. & DELROT S. 2001. The functions of inter- and intracellular glutathione transport systems in plants. - Trends Plant Sci. 6: 486–492.

FRANKEL E. N. 2007. Antioxidants in food and biology – facts and fiction. - The oily press. Bridgewater. England.

GALLARDO-GUERRERO L., GANDUL-ROJAS B., ROCA M. & MÍNGUEZ-MOSQUERA M. I. 2005. Effect of Storage on the Original Pigment Profile of Spanish Virgin Olive Oil. – JAOCS 82(1): 33-39.

GALMÉS J., ABADÍA A., CIFREA J., MEDRANO H. & FLEXAS J. 2007a. Photoprotection processes under water stress and recovery in Mediterranean plants with different growth forms and leaf habits. - Physiol. Plant. 130: 495–510.

GALMÉS J., RIBAS-CARBO M., MEDRANO H. & FLEXAS J. 2007b. Response of leaf respiration to water stress in Mediterranean species with different growth forms. - J. Arid. Environ. 68: 206-222.

GARCÍA-PLAZAOLA J. I. & BECERRIL J. M. 2000. Effects of drought on photoprotective mechanisms in European beech (*Fagus sylvatica* L.) seedlings from different provenances. - Trees 14: 485-490.

GARROTE R. L., SILVA R. & BERTONE R. A. 1986. Losses by diffusion of ascorbic acid during water blanching of potato tissue. - LWT 19(3): 263–265.

GARZÓN G. A. & WROLSTAD R. E. 2009. Major anthocyanins and antioxidant activity of Nasturtium flowers (*Tropaeolum majus*). - Food Chem. 114: 44-49.

GASPAROTTO Jr. A., BOFFO M. A., BOTELHO-LOURENÇO E. L., ALVES-STEFANELLO M. E., KASSUJA C. A. L. & ANDRADE-MARQUES M. C. 2009. Natriuretic and diuretic effects of *Tropaeolum majus* (Tropaeolaceae) in rats. - J. Ethnopharm. 122: 517-522.

GAWRYSIAK-WITULSKA M., SIGER A. & NOGALA-KALUCKA M. 2009. Degradation of tocopherols during near-ambient rapeseed drying. – J. Food Lipids 16(4): 524-539.

GAZZANI G., PAPETTI A., MASSOLINI G. & DAGLIA M. 1998. Anti- and pro-oxidant activity of watersoluble components of some common diet vegetables and the effect of thermal treatment. – J. Agric. Food Chem. 46: 4118–4122.

GIL M. I., FERRERES F. & TOMAS-BARBERAN F. A. 1998. Effect of modified atmosphere packaging on the flavonoids and vitamin C content of minimally processed Swiss chard (*Beta vulgaris* Subspecies *cycla*). - J. Agric. Food Chem. 46: 2007–2012.

GIL M. I., FERRERES F. & TOMÁS-BARBERÁN F. A. 1999. Effect of postharvest storage and processing on the antioxidant constituents (flavonoids and vitamin C) of fresh-cut spinach. - J. Agric. Food Chem. 47: 2213–2217.

GIL-IZQUIERDO A., GIL M. I., CONESA M. A. & FERRERES F. 2001. The effect of storage temperatures on vitamin C and phenolics content of artichoke (*Cynara scolymus* L.) heads. – Innov. Food Sci. Emerging Tech. 2(3): 199–202.

GÖKMEN V., BAHÇECI K. S., SERPEN A. & ACAR J. 2005. Study of lipoxigenase and peroxidase as blanching indicator enzymes in peas: change of enzymes activity, ascorbic acid and chlorophylls during frozen storage. – LWT – Food Sci. Tech. 38: 903–908.

GONÇALVES E. M., CRUZ R. M. S., ABREU M., BRANDÃO T. R. S. & SILVA C. L. M. 2009. Biochemical and colour changes of watercress (*Nasturtium officinale* R. Br.) during freezing and frozen storage. – J. Food Engin. 93: 32–39.

GOOS K. H., ALBRECHT U. & SCHNEIDER B. 2006. Efficacy and safety profile of a herbal drug containing nasturtium herb and horseradish root in acute sinusitis, acute bronchitis and acute urinary tract infection in comparison with other treatments in the daily practice/results of a prospective cohort study. - Arzneimittelforschung 56: 249–257.

GRANADO F., OLMEDILLA B., BLANCO I., & ROJAS-HIDALGO E. 1992. Carotenoid composition in raw and cooked Spanish vegetables. – J. Agric. Food Chem. 40: 2135–2140.

GRASSI G. & MAGNANI F. 2005. Stomatal, mesophyll conductance and biochemical limitations to photosynthesis as affected by drought and leaf ontogeny in ash and oak trees. - Plant Cell. Environ. 28: 834-849.

GRIFFITHS D. W., DEIGHTON N., BIRCH A. N. E., PATRIAN B., BAUR R. & STÄDLER E. 2001. Identification of glucosinolates on the leaf surface of plants from the Cruciferae and other closely related species. - Phytochemistry 57: 693–700.

GROUNEVA I., JAKOB T., WILHELM C. & GOSS R. 2006. Influence of ascorbate and pH on the activity of the diatom xanthophyll cycle-enzyme diadinoxanthin de-epoxidase. - Physiol Plant. 126: 205–211.

GÜLÇIN İ., BÜYÜKOKUROĞLU M. F., OKTAY M. & KÜFREVIOĞLU Ö. İ. 2002. On the in vitro antioxidant properties of melatonin. – J. Pineal Res. 33: 167–171.

GULLUCE M., SOKMEN M., DAFERERA D., AGAR G., OZKAN H. & KARTAL N., POLISSIOU M., SOKMEN A. & SAHIN F. 2003. The in vitro antibacterial, antifungal and antioxidant activities of the essential oil and methanol extracts of herbal parts and callus cultures of *Satureja hortensis* L. - J. Agric. Food Chem. 51: 3958–3965.

GUNAWAN M. I., & BARRINGER S. A. 2000. Green color degradation of blanched broccoli (*Brassica oleracea*) due to acid and microbial growth. – J. Food Proc. Pres. 24: 253–263.

GUTZEIT D., BALEANU G., WINTERHALTER P. & JERZ G. 2008. Vitamin C Content in Sea Buckthorn Berries (*Hippophaë rhamnoides* L. ssp. *rhamnoides*) and Related Products: A Kinetic Study on Storage Stability and the Determination of Processing Effects. – J. Food Sci. 73(9): 615-620.

HAGER A. 1969. Lichtbedingte pH-Erniedrigung in einem Chloroplasten-Kompartiment als Ursache der enzymatischen Violaxanthin-Zeaxanthin-Umwandlung: Beziehung zur Photophosphorylierung. - Planta 89: 224-243.

HAGER A. 1975. Die reversiblen, lichtabhängigen Xanthophyllumwandlungen im Chloroplasten. - Ber. Deutsch. Bot. Ges. 88: 27-44.

HAGER A. & HOLOCHER K. 1994. Localization of the xanthophyll cycle enzyme violaxanthin de-epoxidase within the thylakoid lumen and abolition of its mobility by a (light-dependent) pH decrease. - Planta 192: 581-589.

HAJHASHEMI V., SADRAEI H., GHANNADI A. R. & MOHSENI M. 2000. Antispasmodic and anti-diarrhoeal effect of *Satureja hortensis* L. essential oil. - J. Ethnopharm. 71: 187-192.

HALLIWELL B. & GUTTERIDGE J. M. C. 1999. Free radicals in biology and medicine. - Oxford University Press. Oxford. UK.

HALVORSEN B. L., CARLSEN M. H., PHILLIPS K. M., BØHN S. K., JACOBS D. R. & BLOMHOFF J. R. 2006. Content of redox-active compounds (i.e., antioxidants) in foods consumed in the United States. – Am. J. Clin. Nutr. 84: 95–135.

HALVORSEN B. L., HOLTE K., MYHRSTAD M., BARIKMO I., HVATTUM E., REMBERG S. F., WOLD A. B., HAFFNER K., BAUGEROD H., ANDERSON L., MOSKAUG J., JACOBS D. R. & BLOMHOFF J. R. 2002. A systematic screening of total antioxidants in dietary plants. – J. Nutr. 132: 461–471.

References

HÄNSEL R. & HAAS H. 1984. Therapie mit Phytopharmaka. – Springer. Berlin. Germany.

HARBORNE J. B. 1963. Plant Polyphenols IX – the glycosidic pattern of anthocyanin pigments. - Phytochemistry 2: 85-97.

HAVAUX M. 1998. Carotenoids as membrane stabilizers in chloroplasts. - Trends Plant Sci. 3: 147–151.

HAVAUX M., EYMERY F., PORFIROVA S., REY P. & DÖRMANN P. 2005. Vitamin E protects against photoinhibition and photooxidative stress in *Arabidopsis thaliana*. Plant Cell 17: 3451–3469.

HEATON J. W., YADA R. Y. & MARANGONI A. G. 1996. Discoloration of coleslaw is caused by chlorophyll degradation. – J. Agric. Food Chem. 44: 395–398.

HELANDER I. M., ALAKOMI H. L., LATVA-KALA K., MATTILA- SANDHOLM T., POL I., SMID E. J., GORRIS L. G. M. & VON WRIGHT A. 1998. Characterization of the action of selected essential oil component on gram- negative bacteria. - J. Agric. Food Chem. 46(9): 3590-3595.

HIDALGO A., BRANDOLINI A. & POMPEI C. 2009. Kinetics of tocols degradation during the storage of einkorn (*Triticum monococcum* L. ssp. *monococcum*) and breadwheat (*Triticum aestivum* L. ssp. *aestivum*) flours. - Food Chem. 116: 821-827.

HOF K. H. V. H., WEST C. E., WESTSTRATE J. A. & HAUTVAST J. G. A. J. 2000. Dietary factors that affect the bioavailability of carotenoids. – J. Nutr. 130: 503–506.

HOLST B. & WILLIAMSON G. 2004. A critical review of the bioavailability of glucosinolates and related compounds. – Nat. Prod. Rep. 21: 425–447.

HOPFER M. 2007. Die Auswirkungen unterschiedlicher Bioregulatoren auf den Photosyntheseapparat und auf die antioxidativen Schutzsysteme der Apfelsorte Golden Delicious und der Birnensorte Williams – Master thesis. Graz.

HOSE E., STEUDLE E. & HARTUNG W. 2000. Abscisic acid and hydraulic conductivity of maize roots: a study using cell- and root-pressure probes. - Planta 211: 874-882.

HUMBECK K., RÖMER S. & SENGER H. 1989. Evidence for an essential role of carotenoids in the assembly of an active photosystem II. – Planta 179: 242-250.

HURD T. R., COSTA N. J., DAHM C. C., BEER S.M., BROWN S.T., FILIPOVSKA A. & MURPHY M. P. 2005a. Glutathionylation of mitochondrial proteins. - Antioxid Redox Signal. 7: 999–1010.

HURD T. R., FILIPOVSKA A., COSTA N. J., DAHM C. C. & MURPHY M. P. 2005b. Disulphide formation in mitochondrial protein thiols. – Biochem. Soc. Trans. 33: 1390–1393.

ITO N., HIROSE M., FUKUSHIMA S., TSUDA H., SHIRAI T. & TATEMATSU M. 1986. Studies on antioxidants: Their carcinogenic and modifying effects on chemical carcinogenesis. - Food Chem. Toxicol. 24: 1071–1082.

JAMBOR J. & CZOSNOWSKA E. 2002. Herbal medicines from fresh plants. - Postepy Fitoterapii 8: 2–5.

JIANG Y. & HUANG B. 2000. Effects of Drought or Heat Stress Alone and in Combination on Kentucky Bluegrass. - Crop Sci. 40: 1358–1362.

JIMÉNEZ A., HERNÁNDEZ J. A., DEL RÍO L.A., SEVILLA F. 1997. Evidence for the presence of the ascorbate-glutathione cycle in mitochondria and peroxisomes of pea leaves. - Plant Physiol. 114: 275–284.

JIMÉNEZ A., HERNÁNDEZ J. A., PASTORI G., DEL RÍO L. A. & SEVILLA F. 1998. Role of the ascorbate-glutathione cycle of mitochondria and peroxisomes in the sencescence of pea leaves. - Plant Physiol. 118: 1327–1335.

JUBANY-MARÍ T., PRINSEN E., MUNNÉ-BOSCH S. & ALEGRE L. 2010. The timing of methyl jasmonate, hydrogen peroxide and ascorbate accumulation during water deficit and subsequent recovery in the Mediterranean shrub *Cistus albidus* L. – Environ. Exp. Bot. 69: 47-55.

JUNG S. 2004. Variation in antioxidant metabolism of young and mature leaves of *Arabidopsis thaliana* subjected to drought. - Plant Sci. 166: 459–466.

KAEFER C. M. & MILNER J. A. 2008. The role of herbs and spices in cancer prevention. – J. Nutr. Biochem. 19: 347–361.

KALT W., FORNEY C. F., MARTIN A. & PRIOR R. L. 1999. Antioxidant Capacity, Vitamin C, Phenolics, and Anthocyanins after Fresh Storage of Small Fruits. - J. Agric. Food Chem. 47: 4638–4644.

KAPOOR R. & NAIR H. 2005: Gammalinolenic acid oils. - In: SHAHIDI F. (ed.), Bailey's Industrial Oil and Fat Products, Edible Oil and Fat Products: Specialty Oil and Oil Products. Vol. 3, 6th Edition, pp. 67–120. John Wile & Sons. New York. USA.

KATSUBE N., IWASHITA K., TSUSHIDA T., YAMAKI K. & KOBORI M. 2003. Induction of apoptosis in cancer cells by bilberry (*Vaccinium myrtillus*) and the antocyanins. - J. Agri. Food Chem. 51: 68-75.

KATZER G. 1999. http://www.uni-graz.at/~katzer/engl/Bora_off.html. - last visited May 17th, 2010.

KATZER G. 2007. http://www.uni-graz.at/~katzer/engl/Satu_hor.html. - last visited May 17th, 2010.

KAUR C. & KAPOOR H. C. 2001. Antioxidants in fruits and vegetables – the millennium's health – Review. - Int. J. Food Sci. Tech. 36: 703-725.

KHACHIK F., BEECHER G. R., GOLI M. B. & LUSBY W. R. 1991. Separation, identification, and quantification of carotenoids in fruits, vegetables and human plasma by high performance liquid chromatography. – Pure Appl. Chem. 63(1): 71–80.

KHACHIK F., GOLI M. B., BEECHER G. R., HOLDEN J., LUSBY W. R., TENORIO M. D. & BARRERA M. R. 1992. Effect of food preparation on qualitative and quantitative distribution of major carotenoid constituents of tomatoes and several green vegetables. – J. Agric. Food Chem. 40(3): 390–398.

KIRSCHBAUM M. U. F. 1988. Recovery of photosynthesis from water stress in *Eucalyptus pauciflora* - a process in two stages. - Plant Cell Environ. 11: 685-694.

KOCH K., DOMMISSE A. & BARTHLOTT W. 2006. Chemistry and Crystal Growth of Plant Wax tubules of Lotus (*Nelumbo nucifera*) and Nasturtium (*Tropaeolum majus*) Leaves on Technical Substrates. - Crystal Growth & Design 6(11): 2571-2578.

KOHATA K., HARAGUCHI T., TSUJI M., UJIHARA T. & HORIE H. 2001. Changes in Contents of Chlorophyll Pigments and Chlorophyllase Activity during Manufacturing of Tencha. - NSKKK 48(10): 744-750.

KOLLIST H., MOLDAU H., OKSANEN E. & VAPAAVUORI E. 2001. Ascorbate transport from the apoplast to the symplast in intact leaves. – Physiol. Plant. 113: 377–383.

KOZLOWSKI T. T. & PALLARDY S. G. 2002. Acclimation and adaptive responses of woody plants to environmental stresses. – Bot. Rev. 68(2): 270–334.

KRANNER I. & GRILL D. 1993. Content of low-molecular-weight thiols during the imbibition of pea seeds. – Physiol. Plant. 88:557-562.

LABUZA T. P. 1996. An introduction to active packaging of foods. - Food Tech. 50(4): 68-71.

LAKO J., TRENERRY V. C., WAHLQVIST M., WATTANAPENPAIBOON N., SOTHEESWARAN S. & PREMIER R. 2007. Phytochemical flavonols, carotenoids and the antioxidant properties of a wide selection of Fijian fruit, vegetables and other readily available foods. - Food Chem.101: 1727–1741.

LANGSETH L. 1995. Oxidants, Antioxidants and Disease Prevension. ILSI Europe Concise Monograph Series. ISBN 0-944398-52-9.

LARSEN K.M., ROBY M. R. & STERMITZ F. 1984. Unsaturated pyrrolizidines from borage (*Borago officinalis*), a common garden herb. – J. Nat. Prod. 47: 747–748.

LAVELLI V., FREGAPANE G. & SALVADOR M. D. 2006. Effect of storage on secoiridoid and tocopherol contents and antioxidant activity of monovarietal extra virgin olive oils. - J. Agri. Food Chem. 54: 3002-3007.

LEE S. K. & KADER A. A. 2000. Preharvest and postharvest factors influencing vitamin C content of horticultural crops. – Postharv. Biol. Tech. 20: 207-220.

LEE K.W., HUR, H.J., LEE H.J. & LEE C.Y. 2005 Antiproliferative effects of dietary phenolic substances and hydrogen peroxide. – J. Agric. Food Chem. 53: 1990–1995.

LEFFINGWELL J. C. 2002. Carotenoids as flavor & fragrance precursors, part of our series on aroma materials roduced by carotenoid degradation. - Leffingwell Reports 2(6).

LEI Y., YIN C. & LI C. 2006. Differences in some morphological, physiological, and biochemical responses to drought stress in two contrasting populations of *Populus przewalskii*. - Physiol. Plant. 127: 182–191.

LEŠKOVÁ E., KUBÍKOVÁ J., KOVÁČIKOVÁ E. , KOŠICKÁ M., PORUBSKÁ J. & HOLČÍKOVÁ K. 2006. Vitamin losses: Retention during heat treatment and continual changes expressed by mathematical models – Review. – J. Food Comp. Anal. 19: 252–276.

LI Y., ZHOU Y., WANG Z., SUN X. & TANG K. 2010. Engineering tocopherol biosynthetic pathway in *Arabidopsis* leaves and its effect on antioxidant metabolism. - Plant Sci. 178: 312-320.

LI T., ZHANG M. & WANG S. 2008. Effects of temperature on *Agrocybe chaxingu* quality stored in modified atmosphere packages with silicon gum film windows. - LWT – Food Sci. Tech. 41(6): 965–973.

LI-COR 1998. Using the LI-6400: Portable photosynthesis system, Book 1, part 1, the basics, - LI-COR Inc. Lincoln. Nebraska.

LIMA C. F., VALENTAO P. C. R. , ANDRADE P. B., SEABRA R. M., FERNANDES-FERREIRA M. & PEREIRA-WILSON C. 2007. Water and methanolic extracts of *Salvia officinalis* protect HepG2 cells from t-BHP induced oxidative damage. - Chemico-Biological Interactions 167: 107-115.

LISIEWSKA Z. & KMIECIK W. 1997. Effect of freezing and storage on quality factors in Hamburg and leafy parsley. - Food Chem. 60(4): 633-637.

LISIEWSKA Z., KMIECIK W. & SLUPSKI J. 2004. Contents of chlorophylls and carotenoids in frozen dill: effect of usable part and pre-treatment on the content of chlorophylls and carotenoids in frozen dill (*Anethum graveolens* L.), depending on the time and temperature of storage. – Food Chem. 84: 511-518.

References

LU D., ZHANG M., WANG S., CAI J., ZHOU X. & ZHU C. 2010. Nutritional characterization and changes in quality of *Salicornia bigelovii* Torr. during storage. LWT – Food Sci. Tech. 43: 519-524.

LU S. & Li L. 2008. Carotenoid Metabolism: Biosynthesis, Regulation, and Beyond – Review. J. Integr. Plant Biol. 50(7): 778-785.

LUAN S. 2002. Signalling drought in guard cells. – Plant Cell Environ. 25: 229–237.

LYKKESFELDT J. & MOLLER B. L. 1993. Synthesis of benzylglucosinolate in *Tropaeolum majus* L. (Isothiocyanates as potent enzyme inhibitors) – Plant Physiol. 102: 609–613.

MADSEN H. L. & BERTELSEN G. 1995. Spices as antioxidants. - Trends Food Sci. Tech. 6: 271–277.

MADSEN H. L., SØRENSEN B., SKIBSTED L. H. & BERTELSEN G. 1998. The antioxidative activity of summer savory (*Satureja hortensis* L.) and rosemary (*Rosmarinus officinalis* L.) in dressing stored exposed to light or darkness. - Food Chem. 63(2): 173-180.

MAEDA H., SAKURAGI Y., BRYANT D. A. & DELLAPENNA D. 2005. Tocopherols protect *Synechocystis* sp. Strain PCC 6803 from lipid peroxidation. - Plant Physiol. 138: 1422-1435.

MAHER J. H. 2010. Fundamentals of Phytonutrition. http://www.greenstogo.com/articles/pdf/Fundamentals_of_Phyto nutrition.pdf -last visited May 17th, 2010.

MAO L., QUE F. & WANG G. 2006. Sugar metabolism and involvement of enzymes in sugarcane (*Saccharum officinarum* L.) stems during storage. - Food Chem. 98(2): 338–342.

MARKUS F., DAOOD H. G., KAPITÁNY J. & BIACS P. A. 1999. Change in the carotenoid and antioxidant content of spice red pepper (paprika) as a function of ripening and some technological factors. - J. Agri. Food Chem. 47: 100-107.

MATA A. T., PROENÇA C., FERREIRA A. R., SERRALHEIRO M. L. M.,. NOGUEIRA J. M. F. & ARAÚJO M. E. M. 2007. Antioxidant and antiacetylcholinesterase activities of five plants used as Portuguese food spices. - Food Chem. 103: 778–786.

MAUGHAN S. & FOYER C. H. 2006. Engineering and genetic approaches to modulating the glutathione network in plants. - .Physiol. Plant. 126: 382–397.

MEISTER A. 1988. Glutathione Metabolism and Its Selective Modification – minireview. – J. Biol. Chem. 263(33): 17205-17208.

MHAMDI B., AIDI WANNES W. & MARZOUK B. 2007. Biochemical evaluation of borage (*Borago officinalis*) rosette leaves through their essential oil and fatty acid composition. – Ital. J. Biochem. 56: 176-179.

MIETKIEWSKA E., GIBLIN E. M., WANG S., BARTON D. L., DIRPAUL J., BROST J. M., KATAVIC V. & TAYLOR D. C. 2004. Seed-specific heterologous expression of a nasturtium FAE gene in *Arabidopsis* results in a dramatic increase in the proportion of erucic acid. - Plant Physiol.136: 2665-2675.

MILOS M., MASTELIC J. & JERKOVIC I. 2000. Chemical composition and antioxidant effect of glycosidically bound volatile compounds from oregano (*Origanum vulgare* L.ssp. *hirtum*). - Food Chem. 71: 79–83.

MITTLER R., VANDERAUWERA S., GOLLERY M. & BREUSEGEM F. V. 2004. Abiotic stress series. Reactive oxygen gene network of plants. - Trends Plant. Sci. 9: 490–498.

MIYOSHI N., TAKABAYASHI S., OSAWA T. & NAKAMURA Y. 2004. Benzylisothiocyanate inhibits excessive superoxide generation in inflammatory leukocytes: Implication for prevention against inflammation-related carcinogenesis. – Carcinogenesis 25: 567–575.

MORAN J. F., BECANA M., ITURBE-ORMAETXE I., FRECHILLA S., KLUCAS R. V. & APARICIO-TEJO P. 1994. Drought induces oxidative stress in pea plants. - Planta 194(3): 346–352.

MOREIRA M. R., ROURA S. I. & DELVALLE C. E. 2003. Quality of Swiss chard produced by conventional and organic methods. - LWT – Food Sci. Tech. 36(1): 135–141.

MOSHA T. C., PACE R. D., ADEYEYE S., LASWAI H. S. & MTEBE K. 1997. Effect traditional processing practices on the content of total carotenoid, beta-carotene, alpha-carotene and vitamin A activity of selected Tanzanian vegetables. - Plant Foods Human Nutr. 50: 189–201.

MOUNTNEY G. J. & GOULD W. A. 1988. Low-temperature food preservation. - In: Practical Food Microbiology and Technology, 3rd edition, pp. 112–115. - Van Nostrand Reinhold Company. New York.

MÜLLER E. 2004. 100 Heilpflanzen selbst gezogen. – Leopold Stocker Verlag. Graz, Stuttgart.

MÜLLER M., HERNANDEZ I., ALEGRE L. & MUNNÉ-BOSCH S. 2006. Enhanced α-tocopherol quinone levels and xanthophylls cycle de-epoxidation in rosemary plants exposed to water deficit during a Mediterranean winter. - J. Plant Physiol. 163: 601–606.

MÜLLER-MOULE P., CONKLIN P. L. & NIYOGI K. K. 2002. Ascorbate deficiency can limit violaxanthin de-epoxidase activity in vivo. - Plant Physiol. 128: 970–977.

MUNNÉ-BOSCH S. 2005. The role of α-tocopherol in plant stress tolerance. – J. Plant Physiol. 162: 743–748.

MUNNÉ-BOSCH S. & ALEGRE L. 2000a. Changes in carotenoids, tocopherols and diterpenes during drought and recovery, and the biological significance of chlorophyll loss in *Rosmarinus officinalis* plants, - Planta 210: 925-931.

MUNNÉ-BOSCH S. & ALEGRE L. 2000b. The significance of β-carotene, α-tocopherol and the xanthophyll cycle in droughted *Melissa officinalis* plants. - Aust. J. Plant Physiol. 27(2): 139-146.

MUNNÉ-BOSCH S. & ALEGRE L. 2002. The function of tocopherols and tocotrienols in plants. – Crit. Rev. Plant Sci. 21: 31–57.

MUNNÉ-BOSCH S. & PENUELAS J. 2004. Drought-induced oxidative stress in strawberry tree (*Arbutus unedo* L.) growing in Mediterranean field conditions. - Plant Sci. 166: 1105–1110.

MUNNS R. & SHARP R. E. 1993. Involvement of abscisic acid in controlling plant growth in soils of low water potential. - Aust. J. Plant Physiol. 20(5): 425-437.

NEGI P. S. & ROY S. K. 2004. Changes in β-carotene and ascorbic acid content of fresh amaranth and fenugreek leaves during storage by low cost technique. - Plant Foods Human Nutr. 58(3): 225-230.

NEILL S. J., DESIKAN R. & HANCOCK J. T. 2002. Hydrogen peroxide signaling. – Curr. Opinion Plant Biol. 5: 388–395.

NEILL S., BARROS R., BRIGHT J., DESIKAN R., HANCOCK J., HARRISON J., MORRIS P., RIBEIRO D. & WILSON I. 2008. Nitric oxide, stomatal closure, and abiotic stress. - J. Exp. Bot. 59: 165-176.

NEUMANN P. 2008. Coping mechanisms for crop plants in drought-prone environments. - Ann. Bot. 101(7): 901-907.

NIEMIETZ A., WANDELT K., BARTHLOTT W. & KOCH K. 2009. Thermal evaporation of multi-component waxes and thermally activated formation of nanotubules for superhydrophobic surfaces. – Progr. Org. Coat. 66: 221–227.

NIIZU P. Y. & RODRIGUEZ-AMAYA D. B. 2005. Flowers and leaves of *Tropaeolum majus* L. as rich sources of lutein. – J. Food Sci. 70: 605-609.

NOCTOR G. 2006. Metabolic signalling in defence and stress: the central roles of soluble redox couples. - Plant Cell Environ. 29: 409–425.

NOCTOR G., GOMEZ L., VANACKER H. & FOYER C. H. 2002. Interactions between biosynthesis, compartmentation and transport in the control of glutathione homeostasis and signalling. – J. Exp. Bot. 53: 1283–1304.

NOCTOR G. & FOYER C. H. 1998. Ascorbate and glutathione: keeping active oxygen under control. - Annu. Rev. Plant Physiol. Plant Mol. Biol. 49: 249–279.

NOCTOR G., ARISI A.-C. M., JOUANIN L., KUNERT K. J., RENNENBERG R. & FOYER C. H. 1998. Glutathione: biosynthesis, metabolism and relationship to stress tolerance explored in transformed plants – review. – J. Exp. Bot. 49: 623–647.

NWUFO M. I. 1994. Effect of water stress on the post-harvest quality of two leafy vegetables *Telfairia occidentalis* and *Pterocarpus soyauxii* during storage. – J. Sci. Food Agric. 64: 265–269.

OH M.-M., CAREY E. E. & RAJASHEKAR C. B. 2009. Environmental stresses induce health-promoting phytochemicals in lettuce. - Plant Physiol. Biochem. 47: 578–583.

OKTAY M., GÜLÇIN İ. & KÜFREVIOĞLU Ö. İ. 2003. Determination of in vitro antioxidant activity of fennel (*Foeniculum vulgare*) seed extracts. Lebensmittel-Wissenchaft und Technologie 36: 263–271.

ÖZCAN M.M., ÜNVER A., UÇAR T., ARSLAN D. 2008. Mineral content of some herbs and herbal teas by infusion and decoction. - Food Chem.106: 1120-1127.

PAGTER M., BRAGATO C. & BRIX H. 2005. Tolerance and physiological responses of *Phragmites australis* to water deficit. – Aquat. Bot. 81: 285–299.

PALA M. 1983. Effect of different pretreatments on the quality of deep frozen green beans and carrots. – Int. J. Refrigeration 6: 238–246.

PARENT B., HACHEZ C., REDONDO E., SIMMONEAE T., CHAUMOTNT F. & TARDIEU F. 2009. Drought and abscisic acid effects on aquaporin content translate into changes in hydraulic conductivity and leaf growth rate: a trans-scale approach. - Plant Physiol. 149(4): 2000-2012.

PARIDA A. K., DAS A. B., SANADA Y. & MOHANTY P. 2004. Effects of salinity on biochemical components of the mangrove *Aegiceras corniculatum*. – Aquat. Bot. 80: 77–87.

PERUCKA I. & MATERSKA M. 2007. Antioxidant vitamin contents of *Capsicum annuum* fruit extracts as affected by processing and varietal factors. - Acta Sci. Pol. Technol. Aliment. 6(4): 67–74.

PFEIFHOFER H. 1989. Evidence of chlorophyll b and lack of lutein in *Neottia nidus-avis* plastids. - Biochem. Physiol. Pflanzen 184: 55-61.

PICCIARELLI P. & ALPI A. 1987. Embryo-suspensor of *Tropaeolum majus*: identification of gibberellins A_{63}. - Phytochemistry 26: 329–330.

PICCIARELLI P., ALPI A., PISTELLI L. & SCALET M. 1984. Gibberellin-like activity in suspensors of *Tropaeolum majus* L. and *Cytisus laburnum* L.. - Planta 162: 566–568.

PIGA A., CARO A. D., PINNA I. & AGABBIO M. 2003. Changes in ascorbic acid, polyphenol content and antioxidant activity in minimally processed cactus pear fruits. - LWT 36(2): 257–262.

PINTÃO A. M., PAIS M. S., COLEY H., KELLAND L. R. & JUDSON I. R. 1995. In vitro and in vivo anti-tumor activity of benzylisothiocyanate: a natural product from *Tropaeolum majus*. Planta Med. 61: 233–236.

PINTEA A., BELE C., ANDREI S., & SOCACIU C. 2003. HPLC analysis of carotenoids in four varieties of *Calendula officinalis* L. flowers. - Acta Biol. Szegediensis 47:37-40.

PIZZOCARO F., SENESI E., QUERRO,O. & GASPAROLI A. 1995. Blanching effect on carrots. Study of the lipids stability during the frozen conservation. - Industrie Alimentari 34: 1265–1272.

PLAYER M. E., KIM H. J., LEE H.O. & MIN D. B. 2006. Stability of α,γ, or δ-tocopherol during soybean oil oxidation. – J. Food Sci. 71(8): 456-460.

POLLE A. 2001. Dissecting the superoxide dismutase-ascorbate-glutathione-pathway in chloroplasts by metabolic modeling. Computer simulations as a step towards flux analysis. - Plant Physiol. 126: 445–462.

PRABHU S. & BARRETT D. M. 2009. Effects of storage condition and domestic cooking on the quality and nutrient content of African leafy vegetables (*Cassia tora* and *Corchorus tridens*). - J. Sci. Food Agric. 89: 1709-1721.

PRIOR R. L. & CAO G. 2000. Antioxidant phytochemicals in fruits and vegetables: diet and health implications. – Hort. Sci. 35(4): 588–592.

PUCKETTE M. C., WENG H., MAHALINGAM R. 2007. Physiological and biochemical responses to acute ozone-induced oxidative stress in *Medicago truncatula*. - Plant Physiol. Biochem. 45: 70-79.

RAJU A. M., VARAKUMAR B. S., LAKSHMINARAYANA A. R., KRISHNAKANTHA T. P. & BASKARAN A.V. 2007. Carotenoid composition and vitamin A activity of medicinally important green leafy vegetables. - Food Chem. 101(4): 1598–1605.

REDDY V., UROOJ A. & KUMAR A. 2005. Evaluation of antioxidant activity of some plant extracts and their application in biscuits. - Food Chem. 90: 317–321.

RICE-EVANS C. A., MILLER J. & PAGANGA G. 1997. Antioxidant properties of phenolic compounds - review. - Trends Plant Sci. 2: 152–159.

ROCK C. L., LOVALVO J. L., EMENHISER C., RUFFIN M. T., FLATT S. W. & SCHWARTZ S. J. 1998. Bioavailability of β-carotene is lower in raw than in processed carrots and spinach in women. – J. Nutr. 128: 913–916.

RODRIGO R., GUICHARD C. & CHARLES R. 2007. Clinical pharmacology and therapeutic use of antioxidant vitamins. - Fundam. Clin. Pharmacol. 21(2): 111–127.

RODRIGUEZ E. B. & RODRIGUEZ-AMAYA D. B. 2007. Formation of apocarotenals and epoxy-carotenoids from β-carotene by chemical reactions and by autoxidation in model systems and processed foods. - Food Chem. 101: 563–572.

References

RODRIGUEZ-AMAYA D. B. 2002. Effects of processing and storage on food carotenoids. - Sight Life Newsletter (special issue) 3: 25–35.

RODRIGUEZ-AMAYA D. B. 2009. Enhancing the carotenoid levels of foods through agriculture and food technology. FoodAfrica, Internet Forum 31st March-11th April; http://foodafrica.nri.org/nutrition/internetpapers/DeliaBRodriguez.pdf. last visited June 7th, 2010.

ROURA S. I., DAVIDOVICH L. A. & DELVALLE C.E. 2000. Quality loss in minimally processed Swiss chard related to amount of damaged area. - LWT – Food Sci. Tech. 33(1): 53–59.

SABLIOV C. M., FRONCZEK C., ASTETE C. E., KHACHATURYAN M., KHACHATRYAN L. & LEONARDI C. 2009. Effects of temperature and UV light on degradation of α-tocopherol in free and dissolved form. - J. Am. Oil. Chem. Soc. 86: 895-902.

SAFER A. M. & AL-NUGHAMISH A. J. 1999. Hepatotoxicity induced by the anti-oxidant food additive, butylated hydroxytoluene (BHT), in rats: An electron microscopical study. – Histol. Histopathol. 14(2): 391–406.

SAHIN F., KARAMAN I., GÜLLÜCE M., ÖĞÜTÇÜ H., ŞENGÜL M., ADIGÜZEL A., ÖZTÜRK S. & KOTAN R. 2003. Evaluation of antimicrobial activites of *Satureja hortensis* L.. - J. Ethnopharmacol. 87: 61-65.

SANTO A., MARTINS I., TOMY S. & FERRO V. 2007. Anticoagulant in vitro effect of hidro-ethanolic extract of edible leaves and flowers of *Tropaeolum majus* L. (Tropaeolaceae) on human plasma. - Latin Am. J. Pharm. 26(5): 732–736.

SCHREINER M., KRUMBEIN A., MEWIS I., ULRICHS C. & HUYSKENS-KEIL S. 2009. Short-term and moderate UV-B radiation effects on secondary plant metabolism in different organs of nasturtium (*Tropaeolum majus* L.). - Innov. Food Sci. Emerg. Tech. 10: 93-96.

SCHROEDER J. I., KWAK J. M. & ALLEN G. J. 2001. Guard cell abscisic acid signalling and engineering drought hardiness in plants. – Nature 410: 327–330.

SCHÜLLER J. 2010. Mikroskopische Untersuchungen an Blättern verschiedener Kräuter nach Trockenstresseinfluss. Master thesis. Karl-Franzens-University Graz.

SCHULTZ O.E. & GMELIN R. 1954. Mustard oil glycoside of *Tropaeolum majus* L. (Indian cress) and the relations of mustard oil glycosides to the growth substances. - Arch. Pharm. 287(6): 342–350.

SEEL W. E., HENDRY G. A. F. & LEE J. A. 1992. The combined effect of desiccation and irradiance on mosses from xeric and hydric habitats. - J. Exp. Bot. 43(8): 1023–1030.

SEFIDKON F., ABBASI K. & KHANIKI G. B. 2006. Influence of drying and extraction methods on yield and chemical composition of the essential oil of *Satureja hortensis*. Food Chem. 99: 19-23.

SEFIDKON F., JAMZAD Z. & MIRZA M. 2004. Chemical variation in the essential oil of *Satureja sahendica* from Iran. - Food Chem. 88: 325–328.

SELOTE D. S. & KHANNA-CHOPRA R. 2006. Drought acclimation confers oxidative stress tolerance by inducing co-ordinated antioxidant defense at cellular and subcellular level in leaves of wheat seedlings. - Physiol. Plant. 127: 494–506.

SHAHIDI F., JANITHA P. K. & WANASUNDARA P. D. 1992. Phenolic antioxidants. Crit. Rev. Food Sci. Nutr. 32: 67–103.

ŠIRCELJ H., TAUSZ M., GRILL D. & BATIČ F. 2007. Detecting different levels of drought stress in apple trees (*Malus domestica* Borkh.) with selected biochemical and physiological parameters. - Sci. Hort. 113(4): 362–369.

SMIRNOFF N. & WHEELER G. L. 2000. Ascorbic acid in plants: biosynthesis and function. – Crit. Rev .Plant Sci. 19: 267–290.

SMIRNOFF N. 2000. Ascorbic acid: metabolism and functions of a multi-facetted molecule. – Curr. Opinion Plant Biol. 3(3): 229-235.

SMIRNOFF N., CONKLIN P. L. & LOEWUS F. A. 2001. Biosynthesis of ascorbic acid in plants: A renaissance. Review. - Annu. Rev. Plant Physiol. Plant Mol. Biol. 52: 437–467.

SOFO A., TUZIO A. C., DICHIO B. & XILOYANNIS C. 2005. Influence of water deficit and rewatering on the components of the ascorbate–glutathione cycle in four interspecific *Prunus* hybrids. - Plant Sci. 169: 403–412.

SONG J.-Y., AN G.-H. & KIM C.-J. 2003. Color, texture, nutrient contents, and sensory values of vegetable soybeans [*Glycine max* (L.) Merrill] as affected by blanching. - Food Chem. 83: 69-74.

SRIVALLI B., SHARMA G. & KHANNA-CHOPRA R. 2003. Antioxidative defense system in an upland rice cultivar subjected to increasing intensity of water stress followed by recovery. - Physiol. Plant. 119(4): 503–512.

STICHA K., KENNEY P., BOYSEN G., LIANG H., SU X., WANG, UPADHYAYA P. & HECHT S. S. 2002. Effects of benzyl isothiocyanate and phenethyl isothiocyanate on DNA adduct formation by a mixture of benzo[a]pyreneand4-(methylnitrosamino)-1-(3-pyridyl)-1-butanone in A/J mouse lung. - Carcinogenesis 23: 1433−1439.

SUHAJ M. 2006. Spice antioxidants isolation and their antiradical activity: a review. - J. Food Comp. Anal. 19: 531-537.

SULZBERGER R. 2002. Gartenkräuter. – BLV Verlagsgesellschaft mbH. München.

SZABÓ I., BERGANTINO E. & GIACOMETTI G. M. 2005. Light and oxygenic photosynthesis: energy dissipation as a protection mechanism against photo-oxidation. - EMBO reports 6: 629–634.

SZALAI G., KELLÖS T., GALIBA G. & KOCSY G. 2009. Glutathione as an antioxidant and regulatory molecule in plants under abiotic stress conditions. – J. Plant Growth Regul. 28: 66–80.

TAJKARIMI M. M., IBRAHIM S. A. & CLIVER D. O. 2010. Antimicrobial herb and spice compounds in food - Review. - Food Control 21(9): 1199–1218.

TAKAHAMA U. 2004. Oxidation of vacuolar and apoplastic phenolic substrates by peroxidase: Physiological significance of the oxidation reactions. – Phytochem. Rev. 3: 207–219.
TALALAY P. & FAHEY J. W. 2001. Phytochemicals from cruciferous plants protect against cancer by modulating carcinogen metabolism. – J. Nutr. 131: 3027−3033.

TAUSZ M., KRANNER I. & GRILL D. 1996. Simultaneous determination of ascorbic acid and dehydroascorbic acid in plant materials by high performance liquid chromatography. – Phytochem. Anal. 7: 69-72.

TOIVONEN P. M. A. 1997. The effects of storage temperature, storage duration, hydro-cooling, and micro-perforated wrap on shelflife of broccoli (*Brassica oleracea* L.; Italica Group). – Postharv. Biol. Tech. 10: 59–65.

References

TORRES-JIMENEZ I. B. & QUINTANA-CARDENES I. J. 2004. Comparative analysis on the use of medicinal plants in traditional medicine in Cuba and the Canary Islands. - Revista Cubana de Plantas Medicinales 9(1).

TRABER M. G & ATKINSON J. 2007. Vitamin E, antioxidant and nothing more - Review. - Free Rad. Biol. Med. 43: 4-15.

TREBST A., DEPKA B. & HOLLÄNDER-CZYTKO H. 2002. A specific role for tocopherol and of chemical singlet oxygen quenchers in the maintenance of photosystem II structure and function in *Chlamydomonas reinhardtii*. - FEBS Letters 516: 156–160.

UCKIAH A., GOBURDHUN D. & RUGGOO A. 2009. Vitamin C content during processing and storage of pineapple. – Nutr. Food Sci. 39(4): 398-412.

ULRICH-MERZENICH G., ZEITLER H., VETTER H. & KRAFT K. 2009. Synergy research: Vitamins and secondary plant components in the maintenance of the redox-homeostasis and in cell signaling. - Phytomedicine 16: 2-16.

VAN HOVE L. W. A., BOSSEN M. E., SAN GABINO B. G. & SGREVA C. 2001. The ability of apoplastic ascorbate to protect poplar leaves against ambient ozone concentrations: a quantitative approach. – Environ. Poll. 114: 371–382.

VANACKER H., CARVER T. L. W. & FOYER C. H. 1998. Pathogen-induced changes in the antioxidant status of the apoplast in barley leaves. - Plant Physiol. 117(3): 1103–1114.

VANDERSLICE J. T., HIGGS D. J., HAYES J. M. & BLOCK G. 1990. Ascorbic acid and dehydroascorbic acid content of foods-as-eaten. - J. Food Compos. Anal. 3: 105–118.

VAUGHAN J. G. & JUDD P. A. 2006. The oxford book of health foods - a comprehensive guide to natural remedies. – Oxford University Press Inc. New York.

VERES S., TÓTH V. R., LÁPOSI R., OLÁH V., LAKATOS G. & MÉSZÁROS I. 2006. Carotenoid composition and photochemical activity of four sandy grassland species. - Photosynthetica 44(2): 255-261.

VISHALAKSHI DEVI D. 2003. Influence of heat processing and storage on the stability of antioxidant vitamins in some plant foods. Dissertation. University of Mysore, India.

WANG W., VINOCUR B. & ALTMAN A. 2003. Plant responses to drought, salinity and extreme temperatures: towards genetic engineering for stress tolerance - Review. - Planta 218: 1–14.

WETTASINGHE M., SHAHIDI F., AMAROWICZ R. & ABOU-ZAID M. M. 2001. Phenolic acids in defatted seeds of borage (*Borago officinalis* L.). - Food Chem. 75: 49–56.

WILDI B. & LÜTZ C.1996. Antioxidant composition of selected high alpine plant species from different altitudes. - Plant Cell Environ. 19: 138-146.

WILLCOX J. K., ASH S. L. & CATIGNANI G.L. 2004. Antioxidants and prevention of chronic disease. - Crit. Rev. Food Sci. Nutr. 44(4): 275–295.

WOJDYLO A., OSZMIANSKI J. & CZEMERYS R. 2007. Antioxidant activity of phenolic compounds in 32 selected herbs. - Food Chem. 105: 940–949.

YAGI K. 1987. Lipid peroxides and human disease. – Chem. Phys. Lipids 45: 337–351.

YAMAMOTO H. Y. 1979. Biochemistry of the violaxanthin cycle in higher plants. - Pure Appl. Chem. 51: 639-648.

YAMASAKI K., NAKANO M., KAWAHATA T. MORI H., OTAKE T., UEDA N., OISHI I., INAMI R., YAMANE M., NAKAMURA M., MURATA H. & NAKANISHI T. 1998. Anti-HIV-1 activity of herbs in Labiatae. - Biol. Pharm. Bull. 21(8): 829-833.

YANISHLIEVA N.V., MARINOVA E. & POKORNÝ J. 2006 Natural antioxidants from herbs and spices - Review. - Eur. J. Lipid Sci. Technol. 108: 776–793.

YIN C., DUAN B., WANG X. & LI C. 2004. Morphological and physiological responses of two contrasting poplar species to drought stress and exogenous abscisic acid application. - Plant Sci. 167(5): 1091–1097.

ZANETTI G. D., MANFRON M. P. & HOELZEL S. C. S. 2004. Análise morfo-anatômica de *Tropaeolum majus* L. (Tropaeolaceae). IHERINGIA Série Botânica 59: 173–178.

ZHANG J. & KIRKHAM M. B. 1996. Antioxidant responses to drought in sunflower and sorghum seedlings. - New Phytol. 132: 361–373.

ZHENG W. & WANG S. Y. 2001. Antioxidant activity and phenolic compounds in selected herbs. – J. Agric. Food Chem. 49(11): 5165–5170.

ZIELIŃSKY H., MICHALSKA A., PISKUŁA M. K: & KOZŁOWSKA H. 2006. Antioxidants in thermally treated buckwheat groats. - Mol. Nutr. Food Res. 50: 824-832.

ZUSSMAN J., AHDOUT J. & KIM J. 2010. Vitamins and photoaging: Do scientific data support their use? - Review. - in press. J. Am. Acad. Dermatol., dci: 10.1016/j.jaad.2009.07.037.

Summary

Under biological conditions plants have to cope with the generation of reactive oxygen species (ROS), which often react radically and are therefore responsible for oxidative damage in cells. Normally, the production of ROS is compensated by an elaborate endogenous antioxidant system consisting of two main mechanisms: antioxidant defense with various enzymes or with non-enzymatic components (e.g. **ascorbic acid, tocopherol, glutathione and carotenoids**), but under unfavorable environmental conditions like drought the production of ROS can increase dramatically. As the availability of water is of great importance for plant growth, one focus of this applied work is on the changes of the antioxidative network induced by drought. Therefore, in our studies climate chamber grown **seasoning herbs (*Tropaeolum majus* L., *Borago officinalis* L. and *Satureja hortensis* L.)** are exposed to a mild drought stress which assures that the plant defence network is activated, but there is no damage in the plants. Concentrations of various antioxidants (mentioned above) and **photosynthetic parameters** are measured in plants without stress, after mild **drought** and after **re-watering**.

As there is an increasing interest in spices and aromatic herbs, because of their strong antioxidant and antimicrobial properties, another focus of this applied work is on the changes of antioxidants in herbs during **drying, storage and processing**. In general, herbs can improve flavour, avoid deterioration and enrich foods with vitamins, but the latter are sensitive to e.g. water loss, light or heat and can degrade easily (depending on various factors) which is a major problem for food industry. Therefore, in our studies field grown plant material is dried, stored for months in different types of bags, and processed in different ways before examining contents of antioxidants to simulate industrial conditions and study the degradation processes.

Zusammenfassung

Pflanzen sind permanent verschiedenen Umweltfaktoren (Trockenheit, Frost usw.) ausgesetzt, die in weiterer Folge verschiedenste Stressreaktionen auslösen, wie z.B. den Anstieg der reaktiven Sauerstoff-Spezies (ROS) im Gewebe. Diese ROS reagieren oft in Radikalreaktionen mit Zellbestandteilen wie Membranproteinen und –lipiden, was die Beeinträchtigung des Zellstoffwechsels zur Folge hat. Aus diesem Grund besitzen Pflanzen ein komplexes Netzwerk aus verschiedensten Enzymen und Antioxidantien (z.B. **Ascorbinsäure, Tocopherol, Glutathion** und **Carotinoide**). Ein Schwerpunkt dieser angewandten Dissertation liegt in der Untersuchung von Veränderungen dieser Antioxidantien und der **Photosynthese** aufgrund von Trockenstress. Dazu werden folgende (industriell bedeutende) **Gewürzkräuter** verwendet: *Tropaeolum majus* **L.**, *Borago officinalis* **L.** und *Satureja hortensis* **L.**. Die Pflanzen werden in Klimakammern gezogen und zum Teil **trocken-gestresst** bzw. gestresst und anschließend **wiedergewässert**. Danach werden die Gehalte von den oben genannten Antioxidantien sowie Photosynthese-Parameter bestimmt.

Gewürze und Kräuter sind aufgrund ihrer antioxidativen und antimikrobiellen Wirkung derzeit von großem Interesse für die Lebensmittelindustrie. Deshalb ist ein weiterer Schwerpunkt dieser Arbeit die Untersuchung von Änderungen in den Antioxidantien-Gehalten durch **Trocknungs-, Lagerungs- und Verarbeitungsprozesse**. Generell werden Kräuter zur Geschmacks-verbesserung eingesetzt, sie können aber auch dem Verderben von Gerichten entgegenwirken und diese mit Vitaminen anreichern. Vor allem diese in den Kräutern enthaltenen Vitamine sind aber anfällig für Abbauprozesse, die z.B. durch Trocknung, Hitze und gängige industrielle Prozesse (zur Konservierung) ausgelöst werden können. In unseren Studien werden die (Freiland-)Kräuter deshalb getrocknet, mehrere Monate gelagert und auf verschiedene Weise zubereitet, bevor der verbleibende Gehalt an Vitaminen bestimmt wird.

Appendix

Table I: Drought stress experiments - results of nasturtium. n = 10, s.d. = standard deviation, DW = dry weight, Ci = intercellular CO_2 concentration.

nasturtium	controls		stressed		re-watered	
[μg/g DW]	mean	s.d.	mean	s.d.	mean	s.d.
neoxanthin	307,52	40,41	412,04	43,30	314,00	42,25
violaxanthin	159,82	31,35	155,86	17,73	149,50	25,71
antheraxanthin	40,73	11,23	90,28	32,93	52,53	12,25
lutein+zeaxanthin	1151,24	148,46	1526,24	223,19	1201,18	167,44
chlorophyll b	2496,10	203,13	2761,27	342,37	2212,41	249,31
chlorophyll a	8851,41	815,50	11136,27	1344,08	8425,69	1162,64
alpha-carotene	19,73	9,06	29,11	15,54	25,82	15,66
beta-carotene	767,01	81,44	1001,76	123,25	785,40	102,14
alpha-tocopherol	413,57	111,63	601,15	107,80	806,12	120,79
total ascorbate	7755,48	1019,34	7136,86	1349,08	7518,71	1019,50
chlorophyll a/b [factor]	3,58	0,13	4,05	0,32	3,80	0,35
[nmol/g DW]						
total cysteine	206,45	46,53	246,20	35,15	300,59	83,68
total glutathione	2279,19	475,07	3473,37	1677,12	1364,75	741,47
[% total content]						
cysteine oxidized	74,11	-	57,56	-	68,41	-
cysteine reduced	25,89	-	42,44	-	31,59	-
glutathione oxidized	57,21	-	44,92	-	72,45	-
glutathione reduced	42,79	-	55,08	-	27,55	-
stomatal conductance [mol H_2O m^{-2} s^{-1}]	0,04	0,02	0,01	0,01	0,03	0,01
transpiration rate [mmol H_2O m^{-2} s^{-1}]	1,14	0,37	0,45	0,23	0,66	0,26
assimilation rate [CO_2 m^{-2} s^{-1}]	6,91	2,86	1,04	0,89	5,80	1,44
Ci [μmol CO_2 mol^{-1}]	267,00	29,52	408,91	95,14	119,80	72,58

Table II: Drought stress experiments - results of nasturtium. Changes of assimilation rate, stomatal conductance and transpiration rate during increasing light intensities.

nasturtium	assimilation rate [CO_2 m^{-2} s^{-1}]			stomatal conductance [mol H_2O m^{-2} s^{-1}]			transpiration rate [mmol H_2O m^{-2} s^{-1}]		
light intensity [µmol photons m^{-2} s^{-1}]	controls	stressed	re-watered	controls	stressed	re-watered	controls	stressed	re-watered
1500	5,250	0,496	5,360	0,017	0,005	0,029	0,600	0,147	0,832
1000	4,690	0,026	5,650	0,016	0,002	0,031	0,510	0,052	0,864
800	3,980	-0,558	6,410	0,014	0,001	0,035	0,455	0,037	0,939
500	3,810	-0,385	7,110	0,013	0,001	0,039	0,403	0,030	1,010
350	3,530	-0,679	7,470	0,013	0,001	0,042	0,401	0,037	1,060
200	2,290	-0,546	7,320	0,011	0,001	0,041	0,332	0,040	1,020
100	2,360	-0,061	4,720	0,009	0,002	0,036	0,246	0,058	0,899
50	0,175	-0,087	2,000	0,006	0,002	0,028	0,169	0,054	0,709
0	-1,780	-0,646	-1,180	0,003	0,001	0,023	0,096	0,035	0,574

Table III: Drought stress experiments - results of summer savory and borage. n = 10, s.d. = standard deviation, DW = dry weight.

	summer savory						borage					
	controls		stressed		re-watered		controls		stressed		re-watered	
[µg/g DW]	mean	s.d.	mean	s.d.	mean	s.d.	mean	s.d.	mean	s.d.	mean	s.d.
neoxanthin	204,54	29,18	213,18	29,01	163,39	16,40	198,90	59,93	220,54	58,64	232,53	38,28
violaxathin	168,95	24,98	145,14	30,03	129,56	25,01	318,87	63,51	284,51	98,57	310,93	52,18
antheraxanthin	34,28	6,62	37,20	7,61	29,52	5,14	28,44	7,55	31,22	11,22	30,50	6,48
lutein+zeaxanthin	608,13	71,74	657,79	86,75	538,79	51,57	578,83	157,69	606,09	139,95	646,57	92,68
chlorophyll b	1614,20	207,46	1712,42	232,40	1319,22	125,41	1254,80	324,28	1128,58	253,76	1221,47	187,70
chlorophyll a	5202,98	633,60	5340,86	807,96	4350,75	393,68	4791,09	1082,78	4311,54	772,17	4358,35	757,89
alpha-carotene	6,52	2,58	6,04	5,40	3,00	1,51	2,08	0,58	3,00	1,47	6,39	2,51
beta-carotene	351,74	44,59	375,17	53,77	298,88	31,49	465,99	135,74	440,46	98,22	518,43	69,91
alpha-tocopherol	328,17	26,73	346,50	27,89	380,02	66,02	216,06	59,13	291,39	50,32	282,98	72,55
total ascorbate	1893,55	198,40	2184,44	290,64	2440,25	385,08	1019,92	376,26	924,89	362,80	1470,30	197,39
chlorophyll a/b [factor]	3,23	0,10	3,11	0,11	3,30	0,13	3,87	0,44	3,86	0,32	3,58	0,43
[% total content]												
ascorbate oxidized	40,18	-	44,24	-	15,51	-	66,93	-	74,36	-	56,75	-
ascorbate reduced	59,82	-	55,76	-	84,49	-	33,08	-	25,64	-	43,25	-
[nmol/g DW]												
total glutathione	3560,80	1564,49	3528,60	622,74	2829,00	280,42	4926,80	1718,03	2962,40	989,25	2265,40	509,75
[% total content]												
glutathione oxidized	38,18	-	35,12	-	31,86	-	62,46	-	72,56	-	46,98	-
glutathione reduced	61,82	-	64,88	-	68,14	-	37,54	-	27,44	-	51,03	-

Table IV: Drying and storage experiments - results of nasturtium, summer savory and borage. n = 10, s.d. = standard deviation, n.d. = no detection possible, DW = dry weight.

nasturtium [µg/g DW]	fresh controls mean	s.d.	dried controls mean	s.d.	6 months tie bags stored mean	s.d.	6 months paper bags stored mean	s.d.
total chlorophylls	9926,80	206,01	7213,75	374,15	3700,95	21,74	4333,13	405,58
total carotenoids	2010,63	46,00	1904,49	117,95	800,28	24,05	1060,55	106,17
total tocopherol	266,77	10,04	373,14	18,96	427,19	165,89	366,03	35,96
total ascorbate	14415,48	457,92	5599,09	224,58	n.d.	n.d.	n.d.	n.d.
chlorophyll a/b [factor]	3,51	0,01	2,48	0,02	2,04	0,02	1,89	0,02
total glutathione [nmol/g DW]	3527,39	353,84	7340,42	382,07	5787,55	441,03	5683,85	213,03

summer savory [µg/g DW]	fresh controls mean	s.d.	dried controls mean	s.d.	3 months tie bags stored mean	s.d.	3 months paper bags stored mean	s.d.
total chlorophylls	6119,33	823,46	5278,81	192,80	2344,12	96,67	3603,98	164,41
total carotenoids	1141,99	175,13	956,55	30,63	346,55	10,86	521,29	30,90
total tocopherol	504,78	257,31	177,83	7,36	142,39	5,33	161,51	7,98
total ascorbate	2039,87	221,18	70,10	31,06	n.d.	n.d.	n.d.	n.d.
chlorophyll a/b [factor]	3,16	0,01	2,70	0,00	2,33	0,01	2,39	0,01
total glutathione [nmol/g DW]	1982,75	245,27	524,60	58,43	626,22	58,16	679,59	328,71

borage [µg/g DW]	fresh controls mean	s.d.	dried controls mean	s.d.	3 months tie bags stored mean	s.d.	3 months paper bags stored mean	s.d.
total chlorophylls	8599,45	236,57	7540,94	291,91	6764,77	225,29	3633,00	1606,28
total carotenoids	1817,35	67,29	1304,35	39,34	801,96	18,75	553,47	12,33
total tocopherol	331,59	8,21	115,48	3,99	106,02	12,15	103,76	22,24
total ascorbate	5599,09	224,58	n.d.	n.d.	n.d.	n.d.	n.d.	n.d.
chlorophyll a/b [factor]	3,52	0,02	3,36	0,01	3,05	0,07	3,25	0,00
total glutathione [nmol/g DW]	2252,26	351,64	100,73	16,96	n.d.	n.d.	271,14	34,16

Table V: Processing experiments - results of nasturtium. n = 10, s.d. = standard deviation, DW = dry weight.

nasturtium	fresh controls		dried controls		dried, 5 min. boiled		dried, 20 min. boiled	
[µg/g DW]	mean	s.d.	mean	s.d.	mean	s.d.	mean	s.d.
total chlorophylls	9926,80	206,01	7150,86	353,22	11822,26	1511,70	8428,66	825,16
total carotenoids	2010,63	46,00	1884,39	111,60	2566,99	296,59	2057,16	166,47
total tocopherol	266,77	10,04	366,68	21,90	323,50	12,93	367,12	34,36
total ascorbate	14415,48	457,92	5599,09	224,58	84,27	22,72	156,18	6,95
chlorophyll a/b [factor]	3,51	0,01	2,83	0,53	2,92	0,12	2,64	0,07
total glutathione[nmol/g DW]	3527,39	353,84	7340,42	382,07	1054,67	196,67	808,20	191,37
[% fresh controls]								
pigments	-	-	-24,31	-	20,54	-	-12,16	-
tocopherol	-	-	37,45	-	21,27	-	37,62	-
ascorbate	-	-	-61,16	-	-97,85	-	-98,92	-
glutathione	-	-	108,10	-	-71,57	-	-77,09	-
[% total content]								
glutathione oxidized	50,01	-	90,42	-	75,63	-	69,56	-
glutathione reduced	49,99	-	9,58	-	24,37	-	30,44	-

Table VI: Processing experiments - results of nasturtium. P < 0.05 analyzed by Kruskal-Wallis-ANOVA, n = 10, s.d. = standard deviation, DW = dry weight.

nasturtium	cut material		cut and 5 min. boiled		cut and 20 min. boiled		cut and rested		cut, rested and 5 min. boiled		cut, rested and 20 min. boiled	
[µg/g DW]	mean	s.d.	mean	s.d.	mean	s.d.	mean	s.d.	mean	s.d.	mean	s.d.
total chlorophylls	12176,53	1697,01	15772,75	341,50	12716,49	1269,75	12671,51	1284,07	13228,46	3844,93	10007,26	744,51
total carotenoids	2444,69	325,43	4115,89	74,15	3360,64	314,78	2419,20	258,41	3810,76	105,15	2533,13	189,67
chlorophyll a/b [factor]	3,46	0,09	3,71	0,17	3,45	0,21	3,39	0,06	3,11	1,02	3,43	0,28
total tocopherol	227,56	72,66	475,52	48,87	507,88	73,99	239,34	37,37	401,34	57,19	406,58	29,29
total ascorbate	15961,46	2680,11	1611,50	357,84	2071,94	199,28	15321,72	1896,07	1658,88	367,55	2475,14	659,43
total glutathione [nmol/g DW]	3421,20	1266,55	352,20	97,07	441,60	109,24	3869,20	519,64	327,60	50,65	423,00	117,16
[% fresh controls]												
pigments	22,48	-	66,61	-	34,68	-	26,42	-	42,74	-	5,05	-
tocopherol	-14,70	-	78,25	-	90,38	-	-10,29	-	50,44	-	52,40	-
ascorbate	10,72	-	-88,49	-	-82,83	-	6,29	-	-88,82	-	-85,63	-
glutathione	-3,01	-	-90,02	-	-87,48	-	9,69	-	-90,71	-	-88,01	-
[% total content]												
glutathione oxidized	66,14	-	53,82	-	57,25	-	50,43	-	60,95	-	51,46	-
glutathione reduced	33,86	-	46,18	-	42,75	-	49,57	-	39,05	-	48,54	-

I want morebooks!

Buy your books fast and straightforward online - at one of world's fastest growing online book stores! Environmentally sound due to Print-on-Demand technologies.

Buy your books online at
www.morebooks.shop

Kaufen Sie Ihre Bücher schnell und unkompliziert online – auf einer der am schnellsten wachsenden Buchhandelsplattformen weltweit! Dank Print-On-Demand umwelt- und ressourcenschonend produziert.

Bücher schneller online kaufen
www.morebooks.shop

KS OmniScriptum Publishing
Brivibas gatve 197
LV-1039 Riga, Latvia
Telefax +371 686 204 55

info@omniscriptum.com
www.omniscriptum.com

Printed by Books on Demand GmbH, Norderstedt / Germany